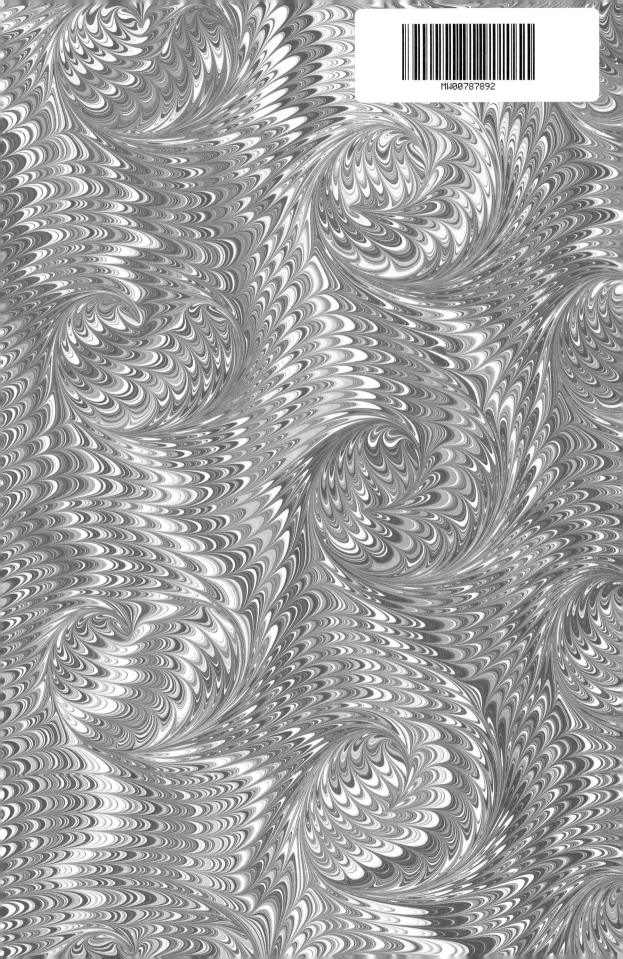

An Illustrated Guide to German Panzers
1935-1945

Wolfgang Fleischer

Series production of Armored Vehicle 1 (MG) (Sd.Kfz. 101), Type A, which first began at the Krupp-Gruson AG in the spring of 1934, 477 vehicles were delivered. The tank was given a multicolored paint job (earth yellow no. 17, brown no. 18, and green no. 28), which could be replaced starting in June 1937 by new coats of dark gray and dark brown paint.

An Illustrated Guide to German Panzers 1935-1945

Werner Regenberg

Schiffer Military History
Atglen, PA

Photo Credits

Anderson (2),
Brunkau (7),
Caye (5),
Doss (1),
Eiermann (24),

Federal Archives (6),
Fischer (8),
Fleischer (182),
Giegling (4),
Hoppe (2),

Klotzsche (3),
Koch (4),
MHM (10),
Thiede (5),
Wetzig (6).

Translated from the German by Ed Force
Book design by Ian Robertson.
Copyright © 2002 by Schiffer Publishing.
Library of Congress Catalog Number: 2001098865

This book was originally published under the title, *Deutsche Panzer 1935-1945, Technik, Gliederung und Einsatzgrundsätze der deutschen Panzertruppe* by Podzun-Pallas.

Printed in China.
ISBN: 0-7643-1556-0

We are always looking for people to write books on new and related subjects. If you have an idea for a book, please contact us at the address below.

Published by Schiffer Publishing Ltd.
4880 Lower Valley Road
Atglen, PA 19310
Phone: (610) 593-1777
FAX: (610) 593-2002
E-mail: Schifferbk@aol.com.
Visit our web site at: www.schifferbooks.com
Please write for a free catalog.
This book may be purchased from the publisher.
Please include $3.95 postage.
Try your bookstore first.

In Europe, Schiffer books are distributed by:
Bushwood Books
6 Marksbury Ave.
Kew Gardens
Surrey TW9 4JF
England
Phone: 44 (0)208 392-8585
FAX: 44 (0)208 392-9876
E-mail: Bushwd@aol.com.
Free postage in the UK. Europe: air mail at cost.
Try your bookstore first.

CONTENTS

A contemporary postcard from a member of an armored division of the Kleist Armored Group on the eastern front, September 1942.

FOREWORD

Only a few subjects have been treated so many times in publications on military history as the history of the German armored troops between 1933 and 1945. Surely the quality of the publications can be evaluated variously. But it is noticeable that the historical military-technological portrayals are far more numerous and thorough. Here, in particular, the numerous publications of W. J. Spielberger must be noted, for scarcely anything better can be found. On the other hand, portrayals of the development of the German armored troops, their tactical basis, their action itself, their successes and failures and the reasons for the latter, have not been the subject of published works as often, even though there are noteworthy works on these themes. They often come, like the book "Erinnerungen eines Soldaten" by General Guderian, from the pens of the protagonists of the German armored troops. Other portrayals remain superficial, have been burdened with subjective cliches, and are thus ill-suited to offer a well-rounded picture of the formation, the service, and the downfall of the German armored troops.

Now, the author of this book has not set the goal of filling in holes that he believes exist. He would just like to make the attempt to focus the interested reader's gaze more sharply on the complexity of the use of armored troops in World War II with a combination of selected first-person reports, reproduced documents, and statistical reports, as well as interesting pictures. Tank warfare was more than the cheerful death-defying deeds of enthusiastic tank soldiers or assaults into the backbones of enemy troops.

Tank warfare was a war of engineers, it was war on the assembly lines, struggling between offense and defense on the battlefield, and lastly, a struggle between quantity and quality. It included thorough training, technical understanding, forward-looking leadership, and lightning-fast situation-based decisions. Tank warfare, though, was also the soldier's everyday life, in this case the tank soldiers between the walls of their "steel coffins."

The author would like to thank Mr. Horst Scheibert, from whom the encouragement to treat this theme came. In addition, Ms. Sonja Wetzig deserves thanks for her thoroughgoing research, as does Mr. Gerhard Thiede for handling the photographic work. Photographic material was made available by, among others, Mr. Caye, Mr. Eiermann, Mr. Koch, and Mr. Brunkau. They, too, deserve thanks.

Freital, Autumn 1995
Wolfgang Fleischer

Panzerkampfwagen I tanks of the motorized troops in the 1935 Nazi Party Day parade in Nuremberg. They were meant to show Germany's regained defensive advantage.

Spain 1936-1939:
Testing Under Fire

Between 1936 and 1939, Spain was the scene of a bloody civil war. On the one side were the Spanish Nationalists under General Franco, with German and Italian support. On the other side was the people's government, which was supported by the Soviet Union. Before this political backdrop the German tanks had their first combat experience.

German technology and German personnel were already on their way to Spain in 1936. The entire undertaking used the camouflage name of "Feuerzauber" (Fire Magic). In September of that year the sending of a complete tank unit had been ordered by Germany. It was given the code name "Drohne" (Drone), and was located at Kubas, a village 40 kilometers from Toledo. Its base was a former carpet factory.

The tank unit was made up of staff, first and second tank companies, transport company, and repair shop. One of the two tank companies was always on duty at the front. Various combat tasks were to be fulfilled there. The second company functioned as a training unit. In October 1936, they began to train Spanish officers, non-commissioned officers, and crews to use Panzerkampfwagen I. The training program was later extended to include training with the 3.7 cm L/45 antitank gun and flamethrowers. To keep up with the increasing personnel needs of the Spanish armored troops—they reached regimental strength in 1939—the "Drohne" tank unit was strengthened by a company during the war. It had the following basic tasks to fulfill:

1 Advising the Spanish forces on the use of tanks and antitank guns. Here the German personnel carried out terrain observation, checked the enemy position, and compared it with their own possibilities of attack. Finally, they took part in combat when needed.

Panzerkampfwagen I (MG) (Sd.Kfz. 101), Type A, of the front company of the "Drohne" tank unit. The unit had the task of accompanying attacks by Spanish tanks. German soldiers participated in the attacks in their tanks after taking part in reconnaissance and planning.

2. The training of new tank crews on German tanks and those of Russian manufacture captured from the People's Army. In addition, training was done on antitank guns, flamethrowers, and transport vehicles. For this, special training camps were established in the spring of 1937.
3. Recovering and repairing their own tanks plus those captured on the battlefield. German soldiers, works masters, and workmen were responsible for this at the repair shop.
4. Gaining and evaluating technical and tactical experience with their own vehicles. It was naturally also important to gain such information from captured equipment. From Kubas, T-26 tanks and BA-10 armored scout cars were sent to the driving school in Kummersdorf, where they were studied thoroughly.

In particular, Germany delivered the Panzerkampfwagen I (MG) (Sd.Kfz. 101), Type A, to Spain. Between 1933 and 1934, 477 of them had been built.

According to information at the Reich War Ministry, Wehrmacht Office, Special Staff W, of November 5, 1937, 91 Panzer I type A tanks had been sent to Spain as of that date. In 1938 an unknown number of Panzer I (MG) (Sd.Kfz. 101), Type B tanks reached the "Drohne" armored unit, plus a few examples of the small Armored Command Car I (Sd.Kfz. 265), Type B. Data as to the total number of tanks sent to Spain vary between 120 and 180. Colonel von Thoma cites 180 tanks for the year of 1938. In comparison, Italy delivered 250 CV 3/33 and CV 3/35 tankettes to the Spanish Nationalists. Soviet Russia let the People's Front government use 362 tanks (primarily T-26, plus a few BT-5) and 100 other armored vehicles.

One of the first attacks was made by tanks of the "Drohne" armored unit on both sides of the road to Villamanta on October 25, 1936. Nineteen vehicles took part. An air attack on the tanks was ineffectual because of the large distances between the vehicles. During the attack, it became possible to capture the

The Spanish infantry practices an attack with tank and flamethrower support in this posed photo for the press.

A damaged tank is towed by a second vehicle, with a small tractor in front.

village of Villamantilla. Further attacks are described by Captain Edler von der Planitz. The rule in all these attacks was that the tanks went before the infantry as a sort of mobile machine-gun nest.

The attack targets seldom extended beyond the tactical framework. Ten to fifteen tanks were used in them; fifty in the course of one attack constituted a rarity. Thus, an attack of the Spanish Nationalists at Guadalajara in which 200 tanks had taken place gained special attention in the press. They could penetrate the enemy defenses up to 50 kilometers, but then had to return. The infantry had not followed, and artillery support was lacking.

Thus, there were essential differences from the conception of armored vehicle use worked out in Germany. Even though the tank was still being developed at that time, and was in conflict with other military theories, a series of features came into dominance: Large tank groups, with the support of motorized artillery, infantry, and engineers, and in conjunction with air forces, were supposed to shatter enemy defenses with mighty plunges into the hinterlands.

The experience gained in the Spanish war was naturally evaluated by all countries, sometimes resulting in false conclusions. One of them said that tanks had lost their value as a means of warfare. In Germany, such false conclusions were greeted with caution. In a report in the journal "Die Kraftfahrkampftruppe" it was decisively stated: "It should be clear to those who understand that, as things seem to be in Spain, in terms of experience as to the value or valuelessness of the tank weapon in and of itself, its significance in relation to other weapons, its purposeful organization, its leadership and use, nothing can be said. It should not be denied that their use in combat there had to show results of a particularly constructive type that could be gained there where fire was heavy from both sides." The 8th Department of the Army General Staff expressed itself very much in this sense in a communique of March 20, 1937. Under the title "Summary of Technical Experiences in Spain," it was first agreed that antitank guns had become serious opponents of the tanks. To conclude from the superiority of the antitank shells to those of the tank guns that antitank forces could be superior to mass attacks of modern tank units, though, was regarded as false. The report went on to state under what conditions tank attacks could be carried out with hopes of success against defense-ready antitank forces:

1. The tank unit itself has to have tanks with guns (3.7 and 7.5 cm) that can put the enemy antitank forces out of action at long ranges, or cripple them with foglaying equipment.

2. The tank attack must be supported by other weapons, especially artillery.

3. The tank attack must be carried out with sufficient forces, particularly with great depth.

Only secondarily, the experts opined, could one counteract the superiority of the antitank shells with heavier armor plate. The weight of heavily armored vehicles and their cost would set limits here, without ultimately guaranteeing victory over the defensive forces.

As for the technical experienced gained with German tanks in Spain, even in view of the special conditions in this theater of war, the result of all considerations was that the Panzerkampfwagen I had not proved itself particularly well. Criticism went to the meager performance of the powerplant of the A version, and the insufficient climbing, clambering, and spanning capability. Hand grenades and dynamite charges could easily penetrate the bottom and top armor. The armor plate in general was too thin, offering hardly any protection from infantry weapons. The main fault, though, was the lack of an effective armor-piercing weapon in the Panzerkampfwagen I. The inferiority to the Russian T-26 tank, which was armed with a 45 cm tank gun, was already noted in reports in the military weekly news in 1937. The "tanque russo" with its precisely firing gun quickly gained the respect it deserved in Spain. Therefore, several Panzer I tanks were fitted with 20 mm Breda guns in Kubas, so as to give them nearly equal firepower—makeshift efforts with the weaknesses of all such efforts.

This tank was repaired by the Interbrigade men and photographed on its first training session in Spain in 1938.

The sum of German experience from using tanks in Spain can be stated in the following principles:

1. The choice of the right offensive tactics was decisive.
2. Tank attacks had to be carried out within the parameters of related weapons (armored divisions).
3. The introduction of tanks with big guns should be hastened.

Not all the lessons learned from using tanks in the Spanish Civil War could be utilized in further development of armored troops in Germany until World War II began. The production capacity of the industry was not sufficient. Of the total 3,396 tanks on hand (as of September 1939), 2,668 were Types I and II. Only 513 tanks were armed with a 3.7 or 7.5 cm tank gun. Of these, 98 were Panzerkampfwagen III (3.7 cm) (Sd.Kfz. 141), 211 Panzerkampfwagen IV (7.5 cm) (Sd.Kfz. 161), and there were 106 Panzerkampfwagen 35(t) (3.7 cm) and 98 Panzerkampfwagen 38(t) (3.7 cm), both Czech-made. There were also 215 armored command cars.

Greater progress was made in the establishment of large armored units. The majority of the tanks were combined in six armored divisions. For every division, a full supply of over 300 vehicles was envisioned; they were to see action in the two tank regiments of the armored brigade. In addition, every division was to have a motorized rifle brigade, an antitank unit, an armored reconnaissance unit with 36 armored scout cars, an artillery regiment with 24 guns, an engineer battalion, an intelligence unit, and the supply troops. Thus, the need for massed tank use along with the related weapons was to be met.

The other side: German Interbrigadesmen with a captured Panzerkampfwagen I (MG) (Sd.Kfz. 101).

Only during the war were Type B Panzerkampfwagen I (MG) (Sd.Kfz. 101) sent to Spain. A few of the small Armored Command Car I (Sd.Kfz. 265) also appeared. Without exception, the command cars were the early type without a cupola. As can be seen in this picture, the ball mantlet for the machine gun is missing.

The superiority of the Russian T-26 impressed the Spanish Nationalists as it did the Germans and Italians. Thus, a few Panzerkampfwagen I tanks were armed with 20 mm Breda Flak guns, in order to be able to hold their own in tank battles. This picture shows such an upgunned tank on the defensive at the Ebro Offensive of the Republicans in July 1938.

Captured tanks were repaired and utilized by both sides in the Spanish Civil War. Here members of the "Drohne" tank unit are transporting a T-26 tank to the workshop in Kubas.

Panzerkampfwagen I (MG) (Sd.Kfz. 101) Type A of the Spanish "Agrupacion des Carros de Combate," seen in a Madrid victory parade in 1939. Captured Russian T-26 B tanks follow in the background.

Supplying the armored divisions with four different basic types depended on development, but was plagued with a series of problems. Let us look first at the question of how the various tank types worked together within a unit (every armored regiment had two units of four tank companies). Thus, serious differences in the fighting strength of Panzerkampfwagen I, II, III, and IV could not be ignored. In the units, the medium tanks armed with tank guns provided the strongest fighting power. Only they were capable of taking on dangerous targets (antitank guns, tanks) quickly and effectively at distances beyond 900 meters. They could support the other tanks, but particularly needed their protection on their flanks. Because of this an attacking formation was developed that could be compared to a steel-core shell. The hard nucleus was formed by the medium Panzer III and IV tanks, the softer shot mantle of the light Panzerkampfwagen I and II. This attacking formation gave the German armored troops their particular nature in the Polish and French campaigns. As the armored units were supplied more and more with medium tanks, they faded into the background.

It is undeniable that their equipping with various tanks, training, command of tanks in combat, and repairs became more complex. These problems were compensated for by outstanding training of the tank crews and commanders at all levels, and by liberal equipping with radios.

In training, much emphasis was placed on the crews' optimal utilization of the terrain to correspond with the nature of their weapon. Heights should be driven around, the smallest low spots should be used to advance under cover, and the same applied to patches of woods and brush.

This picture, taken during prewar maneuvers, illustrates the concept of the "steel-core tactic." Surrounded by light Panzer I and II tanks, a Type A Panzerkampfwagen IV (7.5 cm) (Sd.Kfz. 161) watches over their advance.

When established, the tank companies generally had eight Panzerkampfwagen I (MG) (Sd.Kfz. 101), Type A; in 1936 there were 22. In 1937 the more powerful Panzerkampfwagen I (MG) (Sd.Kfz. 101), Type B, reached the troops.

This picture was taken during a prewar field parade. The use of the death's-head symbol in this form goes back to the German armored troops of World War I days.

The road wheels of the Panzerkampfwagen I consisted of an aluminum wheel with cast-in [Buchsen] and vulcanized rubber tires. Aluminum was a rationed metal and was available to Germany only in small quantities.

A demonstration of massive, unbeatable power, the Panzerkampfwagen I (MG) (Sd.Kfz. 101), Type A, as shown on a contemporary postcard. In the internal evaluations of the Army Weapons Office it did not come off so well: insufficient off-road capability and engine power, too weakly armored and armed.

Panzerkampfwagen I (MG) (Sd.Kfz. 101), Type A, in the sands of Brandenburg. The armored training center at Wünsdorf was of great importance for building up the German armored troops. Here many things were tested that had to prove their worth in the war.

A field parade formation, photographed in the mid-thirties. The multicolored paint, which was meant as "passive air protection" to conceal the contours of the vehicles from aviators' eyes, can be seen clearly. The unit markings on the rear ends of the track aprons are also interesting, being adopted from playing cards.

From October 1937 to March 1938, 35 of the Panzerkampfwagen IV (7.5 cm) (Vs.Kfz. 618, later Sd.Kfz. 161), Type A were built. Although not in desired numbers, they took part in many of the troops' large-scale maneuvers in prewar days. Their crews included drivers from the army test center in Kummersdorf, having been included as members of the armored troops for this purpose.

In 1934 the first armored unit, commanded by Major Harpe, was established. Compared to foreign tanks, the German vehicles, armed only with machine guns, aroused little confidence. Their value for training was undenied. They were inexpensive to produce and allowed large tank units to be formed in a short time.

This tank, photographed in March 1939, skidded into a roadside ditch on the ice and threw its right track.

On the advance into Czechoslovakia, crossing the Erzgebirge proved to be very difficult. Here a Panzerkampfwagen I had to be towed by two Panzerkampfwagen II (2 cm) (Sd.Kfz. 121), Type C, after sliding into a ditch on an icy road.

On the advance into the part of Czechoslovakia populated by the Sudeten Germans, in mid-March 1939, this tank aroused the lively interest of the village boys.

The majority of the almost 900 Panzerkampfwagen I (MG) (Sd.Kfz. 101), Type B tanks were made by the Henschel firm. Without weapons, such a vehicle cost 38,000 Reichsmark. The 3.8 liter Maybach engine helped to improve off-road capability.

In the Panzerkampfwagen II the commander was simultaneously the aiming gunner. The tank's success in battle depended on his capability and firing technique.

Tanks on parade in Nürnberg. The variety of types and variations within one formation is easy to see. In front are Panzerkampfwagen I Types A and B, followed by Panzerkampfwagen II, Types B and C.

Poland 1939:
First Successful Action

The Army utilized five armored divisions in Poland in September 1939. Along with the tanks in the "Kempf Armored Division," a hastily formed special unit of brigade strength (enlarged into the 10th Armored Division after the campaign), and the vehicles used by the light divisions, there were some 2,400 tanks ready for the attack. The experiences gained in the Polish campaign were summed up in a memo about the further technical development of the German tank weapon, dated March 23, 1940: "In the Polish campaign, great success can be credited to the tank weapon in particular where they were able, in the cavalry manner, to make surprise attacks on enemy units, especially marching columns or backline sections. On the other hand, it was seen that the tanks were not up to their actual main task of breaking through fortified fronts in the field (for example, the "Kempf Armored Division" lost 39 tanks in three days at Mlava, seven of them total losses). Their off-road capability was not sufficient to overcome field obstacles, their armor was easily penetrated not only by antitank guns, but also by antitank rifles; and their high speed could not be put to use. The losses of tank weapons through mechanical wear (steady driving on tracks!) and by antitank weapons were quite high.

Their successes were thus attributable first of all to the ability of their crews and capable leadership. Technically, the apparatus was insufficient. If the combat had lasted a few weeks longer, the tank weapon probably would have broken down altogether.

That the armor of the German tank weapon was too weak is shown by a report of a member of the 1st Mountain Division on the combat around Lemberg. On the morning of September 14, 1939, "six light tanks (armed with 2 cm guns and machine guns) met there and drove toward Zboiska for our relief. From there we heard loud sounds of battle a short time later, and later yet we learned that five of these tanks had been disposed of from the rear by Polish antitank gunners."

The Panzerkampfwagen II Type C was first built in 1937. The price per vehicle, without weapons and radio, was about 50,000 Reichsmark. The characteristic running gear included five road wheels with quarter-elliptic springs.

Problems arose in the maintenance of tanks, caused mainly by the lack of war experience. A member of Armored Unit 33 (4th Light Division) recalls: "Their first use was in Poland in 1939. To some extent, heavy shot damage took place here…. As far as possible, defective components were replaced by the repair shop company. Holes that had been made by bullets were welded shut. In one case…a tank had had to take five shots from a very well-camouflaged Polish antitank gun at a railroad crossing. Since the crewmen lay dead inside the tank, the metalworkers refused to get in to begin the work. To remove the crew, the bow plate was taken out. The tank then had to be sent back to Germany as a total loss."

Despite the many faults that turned up, it was ascribable primarily to the use of the tanks that the campaign in Poland was decided militarily within eighteen days. A combination of several armored divisions did not take place. Yet, the message could be sent that the formation of large tank groups gave the military command a decisive weapon. This basic knowledge could not be denied, even by the debacle of the "Kempf Armored Division" on the Mlava (September 1-4, 1939) and the misfortunes of the 4th Armored Division in the streets of Warsaw on September 8, 1939.

In all, 993 armored vehicles were lost in Poland between September 1 and October 31, 1939; 800 of them were tanks. 217 were total losses. They can be broken down as follows:

Type	Total On Hand	Total Losses
Panzerkampfwagen I (MG) (Sd.Kfz. 101)	1445	89
Panzerkampfwagen II (2 cm) (Sd.Kfz. 121)	1223	83
Panzerkampfwagen III(3.7 cm) (Sd.Kfz. 141)	98	26
Panzerkampfwagen 35(t) (3.7 cm)	106	?
Panzerkampfwagen 38(t) (3.7 cm)	98	?
Panzerkampfwagen IV (7.5 cm) (Sd.Kfz. 161)	211	19
Armored Command Car	215	?

The armor protected the two-man crew from fire with SmK ammunition at all ranges and all inclinations of the vehicle up to 30 degrees. The body was not gasproof; two of the Gas Mask 30 were provided for protection against poison gas.

A light tank company with four platoons and a supply train is ready to march out. The company chief had a small Armored Command Car (Sd.Kfz. 265). Three platoons are equipped with Panzerkampfwagen I (Sd.Kfz. 101) of both types, the fourth platoon with Panzerkampfwagen II (2 cm) (Sd.Kfz. 121), Type C.

The first Kampfwagen IV (7.5 cm) (Sd.Kfz. 161), Type C tanks reached the troops in the autumn of 1938. On account of the low level of industrial output and the fast expansion of the armored troops, only a few tanks with tank guns were available.

In Wünsdorf, according to a mechanic from the shops there, interesting experiments were carried out before the war. A turretless Panzerkampfwagen I with aluminum frame and a rolled-up carpet of wooden bars was mentioned, as well as a Panzerkampfwagen III with a flamethrower in place of the turret machine gun.

The higher percentage of total losses of medium tanks in comparison to the numbers used can be attributed to the fact that these vehicles had to be called on more often to carry out combat tasks. In evaluating the experience gained in the Polish campaign, the production of tanks armed with tank guns was particularly hastened.

Still, their production was too meager. That was a handicap for the increase in the number of armored divisions, which was planned to go on until 1940. At first the armored divisions that had been in action were to be filled and ordered, and the light divisions were to be revised into armored divisions. Each one was given two additional tank units, so that with the four new armored divisions (6th, 7th, 8th, and 9th), there could be a total of ten large armored units made ready for the iminent attack on France. The majority of the divisions, with two thirds of all the tanks, became combined in the "Kleist Armored Group," which was intended to be used in a deep, powerful advance to the Channel coast.

In the new organization of the armored groups, the composition of the armored units of three or four companies was retained. Of them, one medium unit was equipped with Panzerkampfwagen IV tanks. The planned number of them for each company was 22 vehicles.

In the spring of 1940, the Panzerkampfwagen III and IV tanks numbered about 600, making one fifth of the total of 3,300 tanks on hand. It is interesting that almost all available Panzer III and IV tanks were intended for action, while only about half of the Panzer I tanks were with the units that were ready for action. An important factor was the addition of numerous 35(t) and 38(t) tanks to turn the light divisions into armored divisions. In May 1940, 143 Panzerkampfwagen 35(t) and 238 Panzer 38 (t) were on hand.

In 1940 the process of equipping a few rifle battalions in the armored divisions with armored personnel carriers began. The first self-propelled artillery guns also made their appearance. Both events took place as part of an effort to utilize the tanks' penetrating power better.

Up to the beginning of the French campaign, the theoretical and practical training of the officers, NCOs, and men was hastened. It required particular efforts to train the required number of tank drivers.

A Panzerkampfwagen II (2 cm) (Sd.Kfz. 121), Type B is seen after the Polish campaign ended. The weapons have been removed because of the heavy dust, and the white cross in the turret has been darkened because it often made targeting easier for the Polish antitank gunners.

The sensitivity of the Panzerkampfwagen III to antitank and field artillery fire had already been noted in the Polish campaign, where in September 1939 a total of 26 tanks of this type had been lost. One of the total losses was this Panzerkampfwagen III (3.7 cm) (Sd.Kfzz. 141)m Type D.

Another picture from the Polish campaign. Such tanks, fully burned out internally, were scrapped, though the specially hardened armor steel was thus lost. In the background are the grave crosses of the five-man crew.

In March 1939, after the occupation of Czechoslovakia, the German Wehrmacht took over 298 Czech LT 35 tanks and designated them 35(t). Of them, 106, plus eight 35(t) Armored Command Cars, saw action in the Polish campaign with the 1st Light Division. This picture shows a damaged 35(t) tank after the Polish campaign.

This Panzerkampfwagen IV (7.5) (Sd.Kfz. 161), Type C, ended up in a junkyard in Poland. The white cross on the driver's armor, 30 mm thick, has been riddled by bullets. The photo was taken in November 1939.

This picture, taken after a drill at the Wünsdorf training center in the summer of 1939, shows that the outer rims of the aluminum road wheels have been worn smooth by driving in sandy soil.

In the first two weeks of the Polish campaign the Army lost 66 Panzerkampfwagen II tanks. Only ten tanks of this type remained in use.

Considerably more modern than the 35(t), and thus more interesting for the speedy training of the armored troops was the Czech LT 38 tank, the production of which by the CKD firm in Prague ended at the close of 1938. The Army took over these tanks under the designation Panzerkampfwagen 38(t). On September 1, 1939, there were 98 of them on hand; they were used by the 3rd Light Division.

The D and E types of Panzerkampfwagen II were intended as fast tanks for the armored units of the light divisions. They reached a top speed of 55 kph. In addition, the light divisions had Faun L 900 D 567 heavy trucks with low loader semi-trailers to transport the tanks. This picture was taken in Poland in September 1939.

The vehicles of this Panzerjäger unit, seen in Poland in September 1939, have to stop at the edge of town to let a tank column go past. At that time the Germans had very little practical experience in the realm of large armored and motorized groups, and what with the stormy offensive, this led to conflicts in the cooperation of the troops.

In all, 42 of the Panzerkampofwagen IV (7.5 cm) (Sd.Kfz, 161), Type B, were delivered. Here a tank of this type, followed by a Repair Vehicle I, passes the 14th (Panzerjäger) company of an infantry regiment in Poland in September 1939.

Structure of the 3rd Armored Division 1939

(simplified portrayal)

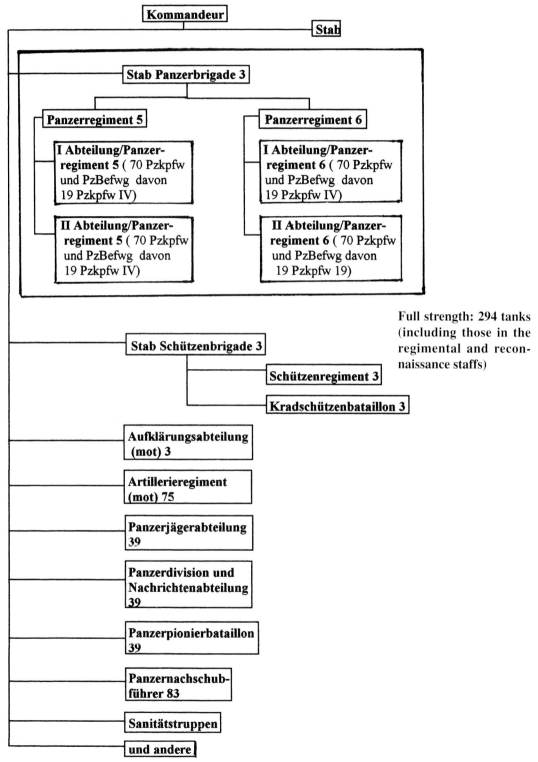

Kommandeur

Stab

Stab Panzerbrigade 3

Panzerregiment 5

I Abteilung/Panzer-regiment 5 (70 Pzkpfw und PzBefwg davon 19 Pzkpfw IV)

II Abteilung/Panzer-regiment 5 (70 Pzkpfw und PzBefwg davon 19 Pzkpfw IV)

Panzerregiment 6

I Abteilung/Panzer-regiment 6 (70 Pzkpfw und PzBefwg davon 19 Pzkpfw IV)

II Abteilung/Panzer-regiment 6 (70 Pzkpfw und PzBefwg davon 19 Pzkpfw 19)

Full strength: 294 tanks (including those in the regimental and recon-naissance staffs)

Stab Schützenbrigade 3

Schützenregiment 3

Kradschützenbataillon 3

Aufklärungsabteilung (mot) 3

Artillerieregiment (mot) 75

Panzerjägerabteilung 39

Panzerdivision und Nachrichtenabteilung 39

Panzerpionierbataillon 39

Panzernachschub-führer 83

Sanitätstruppen

und andere

Vehicles of a division's antitank unit wait beside a road in September 1939 while a Panzerkampfwagen II column passes them. The last tank has a fog-cartridge launcher on its exhaust shield; the fog cartridges are not attached.

In the first days of the Polish campaign, the German tanks still had the very visible white crosses on their turrets and hulls; later they were darkened.

The first version of Panzerkampfwagen IV was not very well protected by its 15 mm armor plate. The ammunition for its 7.5 cm Kampfwagenkanone 37 L/24 consisted of explosive and foglaying shells. The tank with the 75.5 cm gun was officially introduced by order of January 27, 1936, but its production was delayed.

Between September 1938 and August 1939, 134 of the Panzerkampfwagen IV (7.5 cm) (Sd. Kfz. 161) had been produced. There were 211 Panzer IV tanks in action in Poland, the greatest number being Type C models.

Panzerkampfwagen II are seen crossing an 8-ton bridge (Bridge Device B), which could carry up to nine tons. Fully tracked vehicles with weights up to nine tons were regarded as standard loads.

A Panzerkampfwagen II (2 cm) (Sd.Kfz. 121), Type C, seen in a suburb of Warsaw in the morning hours of September 9, 1939. As the history of the tank forces shows, the use of tanks in big cities involved heavy losses. The surprise attack failed; the 4th Armored Division lost 57 of its 120 tanks within three hours.

From 1939 on, about 100 of the Panzerkampfwagen III (3.7 cm) (Sd.Kfz. 141) were built. This version had the final running gear with torsion-bar suspension. This picture was taken in the autumn of 1939, after the Polish campaign, as the units returned to their garrisons.

The German attack on Denmark and Norway in April 1940 was conducted under the code name of "Weserübung." The Germans used mainly Panzerkampfwagen I and II. Often the appearance of a few tanks was enough to decide the battle.

On March 8, 1940, the Tank Unit z.b.V. 40, with three companies, was established for service in Denmark and Norway. The 3./Tank Unit z.b.V. 40, for example, supported Infantry Regiment 236 very effectively in the combat around Haugsbygd on April 16, 1940. The Norwegians were surprised, for they had never yet seen such tanks and had no special defensive weapons.

The so-called "new model vehicles" were developments of heavy multiturreted tanks by the Krupp and Rheinmetall firms. Five of them were built in all. This picture was taken during the assembly of a Rheinmetall tank at the end of the thirties.

Following inspiration from other countries, particularly tank construction in Britain, the Germans began to develop multiturreted tanks in 1932. The vehicles bore various names, but were generally known by the code name of "Neubaufahrzeug." Later they were put to use at the armored troops' firing school in Putlos. Because of the poor stability of their hulls, they showed many cracks. Three "new model vehicles" saw action in Norway in April 1940.

France 1940:
Tank Corps Make the Difference

In the first light of dawn on May 10, 1940, the German attack on Holland, Belgium, and France began. The Kleist Armored Group advanced in the direction of the mouth of the Somme. In the 4th Army, the tanks of the XV. Armored Corps, under General Hoth, rolled toward Dinant. Farther to the north, the XVI Armored Corps of the 6th Army, under General Hoepner, and the 9th Armored Division of the 18th Army attacked. In all, ten armored divisions led the decisive breakthrough to the Channel coast and divided the Allied troops into two halves. The British and French units pushed together in the Dunkerque area had to be evacuated minus their heavy equipment.

On June 5, 1940, the second phase of the Battle of France began. German armored units had been regrouped for it, and now, in close cooperation with infantry divisions, conquered the so-called Weygand Line. They advanced southward on a broad front. Within 20 days the combat had ended successfully, and again the armored troops had played a decisive role in it.

The French and British troops had a very effective defense against tanks, which was supported by a numerically strong artillery. In particular, the 47 mm 37 SA-APX antitank gun was feared. It could penetrate 36 mm of armor plate at a range of 1,500 meters, at a striking angle of 60 degrees. The Panzerkampfwagen III (3.7 cm) (Sd.Kfz. 141)m Type E, could only offer 30 mm of frontal armor against it. The armored troops of the Allies were also superior in numbers. At the beginning of the campaign, 3,354 British and French tanks were on hand at the front (600 of them British).

The 7th Armored Division prepares for an attack in France late in May 1940. At the beginning of the campaign, this division had 48 of the 38(t) tanks on hand.

The first prototypes of the later Panzerkampfwagen III were available at the end of 1935. Of the ten finished vehicles, eight were fitted with the turret developed by Rheinmetall, with a 3.7 cm tank gun. A year later the first three Panzerkampfwagen III (3.7 cm), Type A, reached the 1st Armored Division. They saw service in the Polish campaign. This picture was taken in 1940.

A Panzerkampfwagen III (3.7 cm) (Sd.Kfz. 141), Type E, of Armored Regiment 6 (3rd Armored Division) is seen crossing a K-bridge (24-ton load limit) in May 1940 in the border area between Belgium and France.

In May 1940, preparations were being made for the attack on France. A hundred of the Panzerkampfwagen II (2 cm) (Sd.Kfz. 121) had been produced. Twenty-five each of the previous types, a1 and a2, and 30 of Type a3 had been built. They appeared in France, and a few also served in Russia.

In particular, the French B-1 and S-35 tanks were better armored and armed than the German Panzerkampfwagen III and IV. The reason for the superiority of the German armored troops was their being combined into armored divisions, their better organization, leadership, and training, and the cooperation of the service arms and various forces, as well as the better radio equipment down to the last tank.

Evaluation of the combat experience from the French campaign showed:

1. The armor plate of all German tanks was too weak and, even at long ranges, offered no protection from antitank artillery. That led to serious losses, such as in the 7th Armored Division. In the first phase of the combat in France it lost the equivalent of two tank units. What remained was united into one unit.

2. The armament of the German tanks was too weak, which had led to critical situations and avoidable losses in the numerous battles with French tanks. Even the 7.5 cm tank gun of the Panzerkampfwagen IV could penetrate the armor of the B-1 bis and the S-35 only at long ranges.

3. The mechanical reliability of the German tanks left something to be desired.

In order to help the cause, the armor plate of Types II, III, and IV had to be increased to 60 mm in the near future. The Panzerkampfwagen III became the most important vehicle in the tank unit. The shortage of Panzerkampfwagen III tanks resulted in the use of Panzerkampfwagen 38(t) tanks captured in Czechoslovakia in their place. In addition, the Panzerkampfwagen IV was to function as a support tank.

The development of a new tank gun (5 cm) for the Panzerkampfwagen III had already been begun at the start of 1938. As of 1940 it was regularly installed in the G version, though as a short version with a barrel length of L/42.

The enemy tank in 1940: The 7th Armored Division had a serious situation near Arras on May 25, 1940, when Royal Tank Regiment units attacked with MK II "Matilda" infantry tanks. The armor-piercing shells of the German tank guns bounced off their 80 mm front armor harmlessly.

The French Army had some 600 heavily armored and armed Char B 1 bis tanks at the front on May 10, 1940. The German Panzerkampfwagen III, 349 of which were at the front on the same day, could accomplish nothing against the French front armor with their 3.7 cm tank guns, even at ranges under 1,000 meters.

Faster than the Char B 1 bis, but equally well armed and armored, was the S-35 cavalry tank. 416 of them were supposed to have been on hand and used by the cavalry tank regiments. Tank Regiment 3 (2nd Armored Division) had combat contact with this type of tank on June 16, 1940, and despite clearly recognized direct hits with 3.7 cm antitank shells, they could not shoot down any of them.

May 1940: A Panzerkampfwagen I (MG) (Sd.Kfz. 101), Type B, is guided through a tree obstacle erected by the Belgians. The German advance units had to overcome one obstacle after another in the forest of Ardennes. Here, speed was needed to get the mighty tank thrust to the Channel coast.

This provided a penetrating power of 37 mm at a range of 1,000 meters. After the review of the armament situation in October 1940, 67 Panzerkampfwagen III were delivered in that month with the 3.7 cm and only 15 with the 5 cm tank gun. In the summer of 1941, more efforts were made to rearm the tanks that had 3.7 cm guns with the new 5 cm tank gun. At the same time, the application of added armor plate and strengthening of the running gear were carried out. By May 1941, 161 Panzerkampfwagen III tanks were reequipped at four factories. The whole process, though, moved forward only slowly, for the General Army Office did not have the vehicles to be reequipped ready at the right times.

According to the old principle of increasing the performance of a weapon in use by the troops first by introducing new ammunition, the rearming of the Panzerkampfwagen III was almost paralleled by the introduction of hard-core shells (3.7 and 5 cm), as well as hollow-charge (7.5 cm) shells. In some cases the armor-piercing capability was improved as shown (all data are for ranges from 500 or 400 meters and for a striking angle of 60 degrees for the shell):

Caliber and type of tank gun	Tank shell (explosive)	Tank Shell 40 (hard-core)	Shell 38 HL (hollow charge)
2 cm KWK 30 and 38	14 mm	20 mm	
3.7 cm KWK L/45	29 mm	37 mm (at 400 m)	
3.7 cm KWK 34(t)	29 mm	39 mm	
3.7 cm KWK 38(t)	33 mm	37 mm (at 400 m)	
7.5 cm KWK 37 L/24	38 mm		70 mm

France, June 1940: In all, 523 Panzerkampfwagen I tanks of both types saw action. They were especially numerous in the 3rd and 4th Armored Divisions (109 and 160). Other units, such as the 1st and 7th Armored Divisions, had only 24 and 37 vehicles of this kind.

In France in May 1940, 96 of the small Armored Command Car I (Sd.Kfz. 265) saw service. In the 1st and 4th Armored Divisions, every tank regiment had six of these vehicles. Their distribution in other units is not known.

With this increased performance, combined with a considerable increase in armor, they hoped to be equipped at least for the 1941 war year. Again, an increase in the number of armored divisions was planned, but opposed by the limited production of new tanks. In 1940 895 Panzerkampfwagen III and 280 Panzerkampfwagen IV had been produced. This was not enough. They got by with a further reduction of the specified numbers of tanks in the divisions. The brigade structure was abolished, so that there was just one regiment in every armored division, as a rule with two, or sometimes three, units. After these changes, for example, the 3rd Armored Division had 209 tanks on June 19, 1941 (in September 1939 there had been 324). The greater numbers of Panzerkampfwagen III were supposed to make up for the lower numbers of tanks. In addition, the introduction in the armored units became easier to handle, so that the greater combat power of the new tanks could be utilized more effectively.

For the Balkan operations, ten armored divisions were prepared, as well as several motorized units. The Operation "Marita" showed that large armored groups could be put to effective use, even in the complicated geographical conditions of the Balkan mountains. Often traversing the difficult roads made higher demands on the tanks and their crews than the ensuing combat did. After the operations were finished, the armored divisions were transferred to the border of the Soviet Union.

The five-man crew of a Panzerkampfwagen III (driver, radioman, gunner, loader, and commander) had decisive advantages over the tanks of other countries. The commander did not need to occupy himself with operating the weapons or maintaining radio contact. He could devote himself fully to commanding the tank.

In the 1st Armored Division, troop tests of new tanks had taken place since 1937. Thus, the presence of Types III and IV was higher when the war began. In some companies there were four Panzer I, six Panzer II, seven Panzer III, and one small armored command car on hand.

A Panzerkampfwagen III (3.7 cm) (Sd.Kfz. 141), Type E, of the 1st Armored Division in 1940. On May 10, 1940, there were 249 tanks in the division, 62 of them being Panzer III. An oak leaf was the troop emblem. This platoon leader's tank (531) was in the 5th Company.

Two Panzerkampfwagen III of Tank Regiment 6 are seen in action in France. In front is the 8th Company chief's tank, and behind it a vehicle of the third platoon of that company. In some tank regiments it was customary to give company chiefs multiples of 100 as numbers.

A Panzerkampfwagen III (3.7 cm) (Sd.Kfz. 141), Type E, of the 1st Armored Division, seen in France in June 1940. There were disagreements about the armament of this comparatively large tank, even during its development. The Inspection of Armored Troops had pushed for a 5 cm tank gun, but could not prevail against the Army Weapons Office, which favored a 3.7 cm weapon.

The five-man crew of a Panzerkampfwagen III (the fifth man took the picture). The practical division of duties among the crew members, along with solid training, allowed the effective utilization of the tactical-technical parameters of this combat vehicle.

Two Panzerkampfwagen III (3.7 cm) (Sd.Kfz. 141), Type E, of the 9th Armored Division encountered a French minefield while penetrating the French lines near Amiens, and were badly damaged. The picture was taken in June 1940.

The effect of French antitank weapons on the 30 mm front turret armor of the Panzerkampfwagen III (3.7 cm)(Sd.Kfz. 141), Type E: At this angle, an antitank shell of the 47 mm Antitank Gun mle. 37 (50 mm penetrating power at 500 meters) has struck; below, an antitank shell of the 25 mm Antntank Gun mle. 34 (29 mm penetrating power at 500 meters) has glanced off, and to the right of it, below the roller mantlet, a shell of the same caliber has penetrated.

A fully burned-out Panzerkampfwagen I (MG)(Sd.Kfz. 101), Type B, is seen in a small French town in June 1940. In combat with the French tanks (the smaller vehicles usually had a 37 mm gun), this tank was at a hopeless disadvantage.

Above: *This hard hit in the body of a Panzerkampfwagen II of Tank Regiment 33 was made by a French 47 mm model 37SA antitank gun, and photographed in June 1940.* Below: *Minor shot damage to turrets and hulls was dealt with by the repair-shop companies. Major damage usually caused the vehicle to be sent back home.*

Repairing shot damage was a tedious job. It can be seen here how the hole made by a 25 mm antitank shell on the right side of a Panzerkampfwagen II turret has been patched.

The lack of suitable equipment, according to war strentgh data in the armored units, left many a tank commander searching for makeshift solutions. Here, a Panzerkampfwagen II was turned into a makeshift bridgelayer. This picture comes from the French campaign.

This Panzerkampfwagen II ran afoul of a French mine and was badly damaged. In the foreground are light 1936 model antitank mines that had to be dug out and defused by engineers.

For the French campaign, 955 Panzerkampfwagen II were prepared. Their number per division varied between 15 (8th Armored Division) and 158 (2nd Armored Division). A goodly number of them already had strengthened armor, as can be seen in this picture.

Panzerkampfwagen I (MG) (Sd.Kfz. 101) damaged in the French campaign were modified for other purposes while being repaired, such as for use as driving-school tanks or ammunition carriers. Others were used for testing purposes.

This Panzerkampfwagen II was lost after being hit by French antitank artillery fire. Even after the front armor was strengthened to 30 mm, this type of tank was very sensitive to shot damage. The 25 mm 34 model antitank gun could penetrate the Panzer II effectively from 500 meters on, and the 47 mm 37 model even at 1,500 meters.

Minor repairs and servicing were part of the constant everyday chores of the tank soldiers. This vehicle also shows that fog launchers have been attached.

In April 1940, Heavy Infantry Gun Company 705 (Sfl.) was assigned to the 7th Armored Division and made part of Rifle Regiment 7. In February 1940, 38 such vehicles had been made and were in service with six armored divisions.

The 7th Armored Division reported having 180 tanks on April 12, 1940. Of them, 48 were Type 38(t). Further Czech tanks were assigned to the three units of tank regiments until the French campaign began on May 10, 1940.

Panzerkampfwagen 38(t), Type B, with 25 mm curved front armor. The 3.7 cm Tank Gun 38(t) L/42 fired revised 0.815 kg Czech armor-piercing shells with an initial velocity of 750 m/sec. At 100 meters they could penetrate 41 mm, and at 1,000 meters 27 mm of armor plate (at an angle of 60 degrees).

Training on the Panzerkamphwagen 38(t) in the winter of 1939-40. By May 1940 the number was raised to 238 of these tanks, 228 of which saw service with the 7th and 8th Armored Divisions.

Compared to the conditions in German tanks, the space in the 38(t) was very meager. On the other hand, they were highly praised for their mechanical reliability and off-road capability.

Advancing toward France late in May 1940, powered by a 128 HP engine that consumed 92 liters of fuel in 100 kilometers, giving a possible range of 250 kilometers.

The 7th Armored Division still had five Panzerkampfwagen IV, 31 Panzerkampfwagen II, and 50 Czech 38(t) tanks on May 29, 1940, three weeks after the campaign against France began. The high percentage of usable Czech tanks is noteworthy.

For the Panzerkampfwagen III (3.7 cm) (Sd.Kfz. 141), Type E, no special details were available. Because of censorship, important details, especially of the weapon mounts, had to be retouched for publication in the press.

A Panzerkampfwagen IV of the 3rd Armored Division is being prepared for action in the French campaign in 1940.

A shot-down Panzerkampf-wagen IV (7.5)(Sd.Kfz. 161), Type C, is seen near Stonne, where it took part in heavy and costly fighting between the the French 3rd Armored Division (3e DCR) and the German 10th Armored Division.

As a rule, damaged tanks were brought in by the repair units after a battle and loaded onto Sd.Anh. 116 trailers for transport to repair shops. If the troops' means were insufficient to repair a tank, it was sent to repair facilities in Germany.

This Panzerkampfwagen III (5 cm)(Sd.Kfz. 141), Type G has the newly introduced 5 cm KWK L/42 tank gun. With this weapon, its battle supply was cut from 120 to 99 shells. In addition, the second turret machine gun was eliminated. The external roller mantlet offered better protection from enemy shells.

A Panzerkampfwagen IV (7.5 cm)(Sd.Kfz. 161), Type B or C, with its driver's visor shot away, photographed in the 1940 French campaign. The vehicle belonged to an 8th (medium) company.

The effect of the short-barreled 7.5 cm tank gun, with which the Panzerkampfwagen IV was equipped until March 1942, remained scant against armored targets. With the 6.8 kg red antitank shell, it could penetrate 41 mm of armor plate at a range of 100 meters. At 2,000 meters it could still pierce 30 mm. Only the introduction of hollow-charge shells improved its performance.

To protect the vulnerable tanks, they were transported long distances by rail whenever possible. This picture was taken after the end of the French campaign and shows Panzerkampfwagen IV, Types C and D tanks, with one Type A in the center.

The German manner of warfare used in France in May and June 1940 surprised contemporaries. The former British Prime Minister Lloyd George said: "A very remarkable war! Over on the German side there are no men...just machines! Tanks, armored vehicles, trucks, motorcycles.... There never was such a thing before. This war is very different from the last one."

To keep the advance of armored and motorized vehicles flowing in France, what with its many rivers and canals, required a liberal use of bridges and other crossing techniques.

This picture, taken in the winter of 1941-42, shows a Panzerkampfwagen III (5 cm)(Sd.Kfz. 141), Type F of the 14th Armored Division being used for training.

Panzerkampfwagen III tanks, seen in the winter of 1940-41, are already using the new 5 cm L/42 tank gun in practicing an attack. In the center of the picture is an Armored Command Car III.

Early in April 1941, the vehicles of the 9th Armored Division are being unloaded at a Bulgarian railroad station near the Yugoslav border. On this Panzerkampfwagen III (3.7 cm)(Sd.Kfz. 141), Type E, the division symbol of this unit (at left near the driver's visor), first used in the Balkan campaign, can be seen.

A Panzerkampfwagen I (MG)(Sd.Kfz. 101), Type B, of the 8th Armored Division is seen crossing the Danube on a floating bridge early in March 1941.

In Bulgaria, in the spring of 1941, a Panzerkampfwagen III (5 cm)(Sd.Kfz. 141), Type F, is seen in a German and Bulgarian military parade.

Typical of the appearance of armored units in the spring and summer of 1941 was their mixed equipping with Panzerkampfwagen III that still used the 3.7 cm gun, but there were alrwady some fitted with the 5 cm gun. Here, a column lines up along a railroad line to use it as a road through the Greek mountains, which were simply not passable for tanks.

The quick advance of large German armored groups in central Greece took the Greeks and British by surprise. Here, a Panzerkampfwagen III wades in the bed of the Pinios river for lack of suitable roads.

NEUE IZ

ILLUSTRIERTE ZEITUNG
Berlin, den 6. Mai 1941
Nr. 18 XVII. Jahrgang Preis 20 Pf.

Italien 2 Lire, Jugoslawien 4 Dinar, Frankreich 4 Francs,
Schweiz 40 Rappen, Schweden 48 Öre, Niederlande
20 Cent, Bulgarien 8 Lewa

Durch die griechischen Gebirgsflüsse
Auch an der Südostfront gibt es für unsere Panzer-
waffe keine unüberwindlichen Hindernisse. Welche
Schwierigkeiten sich dem unaufhaltsamen Vor-
marsch entgegenstellten, zeigen die Bilder auf den
Innenseiten.   Aufnahme: PK-Dick /P. J. Hoffmann

After the tanks had won their place as a decisive means of warfare, they also adorned more and more front pages of German magazines.

A Panzerkampfwagen II (2 cm)(Sd.Kfz. 121), Type A-C, serves as a command tank in a light company completely equipped with Panzer I tanks in the Balkans in the spring of 1941.

Panzerkampfwagen II (2 cm)(Sd.Kfz. 121), Type A-CV, of the 9th Armored Division on the march to the Yugoslav border on April 6, 1941. The strengthened armor on the front of hull and turret is easy to see.

On June 1, 1941, the Army had 1,072 Panzerkampfwagen II tanks. In that year 698 of them were built new, while 393 tanks were total losses.

The Type E Panzerkampfweagen IV, made since September 1940, is externally recognizable by the new commander's cupola, the veltilator on the turret roof, and the modified driver's visor. This picture was taken in Greece in the spring of 1941. The first tank already has the newly strengthened upper body armor, which the last tank in the column lacks.

A tank damaged by a mine is being taken away. For this purpose, the rescue column used the 18-ton heavy tractor (Sd.Kfz. 9) with low-loader trailer for tanks (formerly Vs.Ah. 642) with a maximum load of 20,000 kilograms. In addition, 8-ton medium tractor (Sd.Kfz. 7) with small low-loader trailer was available.

This Panzerkampfwagen IV (7.5 cm)(Sd.Kfz. 161), Type D, already has the standard heavier armor on the front and sides of the upper hull. From the Type D on, new tracks, not suitable for the older models, were used.

An Armored Command Car III (Sd/Kfz. 266, 267, 268), Type H is seen crossing the Danube near Giurgiu early in March 1941. The bridge built especially for tanks was 1,100 meters long and designed to carry 24 tons.

This Armored Command Car III, named "Kastor," belonged to the 9th Armored Division. The picture was taken early in March 1941, just before the beginning of Operation "Marita." The dummy tank gun and machine gun in the turret can be seen. The machine gun intended for self-defense and mounted in a ball mantlet was draped with a dust cover.

Pictures for comparison: The tank repair shop of Tank Regiment 33 (9th Armored Division) in France in June 1940 (above), and the same unit barely a year later, in the spring of 1941, doing repair work near Skolpje (below).

The impact of an armor-piercing shell from a Yugoslav field gun has broken out a large piece of the bow armor of this Panzerkampfwagen III. It became ever clearer that the armor of this type was too weak and urgently needed strengthening. The picture was taken in April 1941 in the 9th Armored Division.

In North Africa the Army's armored troops saw action with the 5th Light Division (later 21st Armored Division) and the 15th Armored Division as of the spring of 1941. The conditions in this region (heat, dust, and sand) caused special problems for the vehicles and their crews. The I./Tank Regiment 5 lost 24 of their 32 usable tanks in a three-day march from Agedabia to southeast of el Mechili (320 km) on April 4-7, 1941. Repairs and maintenance burdened the repair shop greatly.

A Panzerkampfwagen III (5 cm)(Sd.Kfz. 141), Type E, of Tank Regiment 5, which had to carry fuel for the march to el Mechili in 20-liter canisters on the vehicle. The 5 cm tank gun was even then no longer enough to meet the demands of tank combat. The British "Matilda" Infantry Tank MK II could only be attacked effectively at ranges under 500 meters, a disadvantage with the usual long ranges.

Tank Regiment 5 of the 5th Light Division was in a preparation area near Gasr el Arid in North Africa on November 17, 1941. They had 124 vehicles ready for use, including 35 Panzerkampfwagen II, 68 Panzerkampfwagen III, and 17 Panzerkampfwagen IV, plus four large armored command cars.

Eastern Front, 1941-1942:
Success and Limits

In June 1941 there were 17 armored divisions in the East, divided along with motorized infantry divisions into four armored groups:

Armored Group 1: Generaloberst von Kleist, with 750 tanks;

Armored Group 2: Generaloberst Guderian, with 930 tanks;

Armored Group 3: Generaloberst Hoth, with 840 tanks;

Armored Group 4: Generaloberst Hoepner, with 570 tanks.

Of the 5,639 tanks on hand in the East on June 1, 1941, 3,300 were concentrated for Operation "Barbarossa." They numbered as follows:

Panzerkampfwagen I (MG)(Sd.Kfz. 101): 180 of the total 877;

Panzerkampfwagen II (2 cm)(Sd.Kfz. 121): 746 of 1,072;

Panzerkampfwagen 35(t)(3.7 cm): ? of 187;

Panzerkampfwagen 38(t)(3.7 cm): 754 of 772;

Panzerkampfwagen III (3.7 and 5 cm)(Sd.Kfz. 141): 965 of 1,440;

Panzerkampfwagen IV (7.5 cm)(Sd.Kfz. 161): 439 of 517;

Armored Command Cars: 230 of 330.

There were also 250 assault guns and an unknown number of captured tanks.

The high percentages of the total numbers of Panzerkampfwagen III and IV is also noticeable. Panzerkampfwagen I had lost much of its significance, and only something more than half of the Panzerkampfwagen II were ready.

The development of new tank models (VK 2001 and VK 3001) was not carried on with any great urgency, what with the success gained with the tanks on hand in the armored units. In 1941 the emphasis was generally

June 1941: The typical battlefield scene in the East. The tanks and riflemen were lost in the wide open spaces of Russia.

on supplying the units with Panzerkampfwagen III and IV. With that, the industry was practically at the limits of its production capacity at that time. In January 1941, 180 tanks were accepted by the Army Weapons Office (88 of them Panzerkampfwagen III and 31 Panzerkampfwagen IV). In September of the same year, the total was 353 vehicles, 206 of them Panzerkampfwagen III and 52 Panzerkampfwagen IV.

It had been seen that Panzerkampfwagen III and IV at first performed above what was required to achieve an increase in combat value. Already carried-out measures to improve firepower (5 cm tank guns for Panzerkampfwagen III) and strengthened armor remained without general disadvantages for tactical and operative mobility.

What conception did the Wehrmacht leadership have of the strength and fighting power of the Red Army, especially its armored troops and artillery? On the one hand, it was incomplete, on the other it was characterized by wishful thinking. The following quotation from a talk between Hitler and Guderian early in August at Novy Borisov, the headquarters of Army Group Center, is significant: Hitler: "If I had known that the number of Russian tanks that you had cited in your book was actually correct, I would (I believe) not have started this war." In his 1937 book "Achtung Panzer," Guderian had stated the number of tanks on hand in Russia at 10,000, thus arousing the disagreement of Chief Beck of the General Staff—and the censors. Now it proved that Guderian had been, if anything, too cautious. The Wehrmacht High Command had already reported on August 6, 1941, that up to that date 13,145 Red Army tanks had been destroyed or captured. Does that surprise anyone? According to Russian sources, the defense industry had produced over 33,000 armored vehicles of all kinds between 1930 and 1941; more than 20,000 of them were tanks. It must be noted that 1,861 of them were the most modern types, the T-34 and KW.

In comparison, on June 22, 1941, the Wehrmacht had 1,404 Panzerkampfwagen III and IV on hand in the East. It is questionable, in view of such statistics, where the often-cited quantitative superiority on the German side comes from.

Despite all the shortages, and despite the quantitative and, in part, also qualitative inferiority of the purely technical components, the Wehrmacht, at the beginning of the Russian campaign, had around 3,300 tanks organized in armored groups, giving it a superior power potential. The following reasons, among others, could be cited for it:

1. The structure and organization of the German armored troops was purposeful and had been able to prove their efficiency in several campaigns. Shortages that had appeared at the beginning were able to be dealt with.
2. The whole technical equipment was designed for lightning-fast and mobile use in the large armored groups. This was not limited to the tanks and their being well equipped with means of communication.
3. The cooperation with the other branches, especially the Luftwaffe, was a firm component of action planning.
4. The enlisted men, NCOs, and officer corps had thorough technical and tactical training.
5. In the previous successful campaigns, valuable experience had been gained and had become the general property of the armored troops.

In all, a significantly more effective utilization of technical and tactical parameters of the available tank technology resulted than was possible in the following war months on the Red Army's side.

To be able to evaluate the German armored troops, it seems to be practical, after the disagreement with the actual war conditions in June 1941, to take up the question of the moment of surprise. This very question is raised again and again to explain the remarkable success of the Germans in the East.

A Panzerkampfwagen III (5 cm)(Sd.Kfz. 141), Type H, of the II/Tank Regiment 18, seen in Russia in June 1941. It originally was a diving tank and belonged to Diving Tank Unit B, which had been assembled in Putlos for Operation "Sealion." After this operation had been called off, the unit was used as the nucleus of the 18th Armored Division. Diving was still done, though, through the Bug on June 22, 1941. This explains the frame for the rubber cover on the roller mantlet and the radioman's machine gun. The diving snorkel is at right on the track apron.

Road conditions in the East made heavy demands on the engineers, especially the armored engineers. Here, a Panzerkampfwagen III of the 14th Armored Division crosses a swampy river bottom near Luzk-Rovno on a bridge made of Bridge Equipment B components.

No less a personage than the Marshal of the Soviet Union, Georgi Zhukov, said after the war: "We had information according to which significant (German) forces were dislocated in Poland.... The present-day explanation of the surprise, but also that which Stalin interpreted in his speeches, is false and incomplete. What is surprise when we talk of actions of that scope? The sudden crossing of the border was in and of itself by no means decisive. The main danger consisted of not being convinced of the striking power of the German Army."

At the first light of dawn on June 22, 1941, the German troops began their attack. With great power the four armored groups penetrated into the depths of the Red Army's defenses. Even in the border combat, the Russian troops suffered heavy losses, even though for the sake of strong resistance. Their centers of resistance, such as at Dubno, were surrounded and isolated.

The toughness of the fighting is reflected in the considerably higher losses in tanks in comparison with the earlier campaigns. In the 3rd Armored Division, one of the strongest combat units, the statistics are:

Numbers of available tanks

June 19, 1941	August 1, 1941	Percentage lost
58 Panzerkampfwagen II	37 Panzerkampfwagen II	36.2%
108 Panzerkampfwagen III	42 Panzerkampfwagen III	61.2 %
32 Panzerkampfwagen IV	16 Panzerkampfwagen IV	50.0%
13 Armored Command Cars	8 Armored Command Cars	38.46%
209 Tanks	103 Tanks	50/7%

Panzerkampfwagen 38(t) and vehicles of the Panzerjäger unit of the 20th Armored Division roll toward Vitebsk on July 11, 1941. In this area there was back-and-forth fighting with Russian T-34 and KW tanks, to which the 38(t) with its 3.7 cm gun was clearly inferior.

This Panzerkampfwagen 38(t), Type B, of Tank Regiment 21 (20th Armored Division) was hit hard by an artillery shell which tore the turret off. The picture was taken near Vitebsk in mid-July 1941.

Damaged tanks were generally assembled by the troops at a central location, from which they were transported back to Germany. This picture shows such a gathering place in the central sector of the eastern front in the summer of 1941. Panzerkampfwagen II were usually used for special tasks on the eastern front because of their weak armament and armor.

The numerous losses of tanks were only attributable in part to the field cannons ("Ratsch-bumm"), of which the Russian divisions had so many. What contributed to the great surprise on the German side was the appearance of a new generation of Russian tanks—the T-34 medium and KW heavy tanks. They embodied, at least in the T-34, a new quality in the combination of mobility, firepower, and armor protection. How had this come about? The Russians had gained very similar experiences to the Germans in the Spanish Civil War. This also concerned the effectiveness of modern antitank guns (3.7 cm caliber), as opposed to the tanks that, until then, had been safe only from machine guns. The Russian designers, of course, emphasized different points in evaluating this experience. For the development of a new generation of tanks, the following criteria became decisive:

1. Effective armor protection against fire from 3.7 cm antitank guns at any range.
2. Increased firepower, in order to put enemy antitank defense (and enemy tanks) out of commission at long ranges
3. Despite the expected increase in weight, the great tactical and operative mobility should be retained or even improved.

A further conclusion was the disbanding of large mechanized groups. That was a mistake. Only the evaluation of the German success of 1939-40 resulted in mechanized corps being reestablished.

Results of efforts in the military realm were the T-34 and the KW. A report of Generaloberst Halder of June 12, 1941, indicates what problems the German antitank forces would face from then on: "...the heaviest tanks were shot down by 10 cm cannons, less often by 8.8 cm flak guns. Light field howitzers with antitank shells also shot down 50-ton tanks at 40 meters...." No German tank had an 8.8 cm or 10 cm tank gun.

An impression of the performance deficit of the German tanks is provided by the following table, which is also relevant to the antitank forces:

Armor thickness/ angle on T-34 Built in 1941	Penetrating power distance	Penetrating power of 3.7 cm KWK L/45 in Pzkw. III	Penetrating power of 5 cm KWK L/42 in Pzkw. III	Penetrating power of 7.5 cm KWK L/24 in Pzkw. IV
Hull, bow 45 mm, 30 degrees	= 90 mm	500 m, 60 degree 29 mm armor	500 m, 60 degree 47 mm armor	500 m, 60 degree 38 mm armor

Note: All data for the German tank guns refer to the fire of explosive tank shells. If one compares the penetrating power of the Russian tank guns with the usual armor thicknesses of Panzerkampfwagen III and IV, a similar overall impression is given for the German side.

Armor thickness/ angle on Pzkw. III H	Penetrating distance	Armor thickness/ angle on Pzkw. IV F1	Penetrating distance	Penetrating performance F-34 L/41 of T-34
Hull, driver's area 30+30 mm, 81 degrees	= 63 mm	Hull, driver's area 50 mm, 80 degrees	=52.5 mm	500 m, 60 degree 75 mm armor

The tanks' enemies: After the eastern campaign began on June 22, 1941, reports on the essentially ineffective armament of tank and antitank troops piled up, for the Russian T-34 tank, of which 53,000 had been built, had appeared.

Another enemy of the German tanks was the WK heavy tank, which was more heavily armored. The German tanks had to be accompanied by 8.8 cm anti-aircraft guns and 10 cm cannons to be able to fight off Russian tank attacks effectively. They had 75 to 100 mm of frontal armor!

For the German tank crews this had fatal consequences:

With the Panzerkampfwagen III (5 cm L/42)(Sd.Kfz. 141), it was only possible to take on the T-34 successfully at distances under 100 meters, and then the new hard-core ammunition had to be available (at 100 meters, striking at a 60-degree angle, it penetrated 94 mm of armor plate). The situation with Panzerkampfwagen IV (7.5 cm L/24)(Sd.Kfz. 161) was somewhat more favorable, as hollow-charge ammunition had been introduced for its 7.5 cm weapon. These shells, though, were available in noteworthy quantities only in the winter of 1941. Their crews had to reckon on being shot down by the T-34's tank gun at ranges over 500 meters. One more disadvantage of this tank was the meager muzzle velocity of its gun, with a meager possibility of hitting moving targets. The results were obvious. Within a short time the German armored divisions lost 50% of their supply of tanks in the East. Of them, 20% had to be written off as total losses, as Halder noted in his war diary in July 1941.

And there were further difficulties:

In a report on the state of the fast units on July 25, 1941, a ten-day refreshing pause was requested for Armored Groups 2 and 3. Particular problems were caused by the above-average technical breakdowns of tanks. The engines suffered from the heavy dust.

On account of the great size of the operation area and the bad roads, the supplying of spare parts became difficult. Early in August 1941, and only after lengthy discussions, did Hitler assure the release of 300 tank engines for the entier eastern front. That was too few. New tanks were not allowed at all. The High Command held them back for setting up new armored units in Germany. For that reason there was particular interest in increasing the production of complete tanks. This was done at the expense of producing spare parts. The shortage of spare parts for repairing damaged tanks grew into a general problem during the war. The lack of stability in tank construction caused another problem. The availability of at least three basic types in the armored units hindered the dismantling of damaged tanks for spare parts, for example, the road wheels of the Panzerkampfwagen II, III, and IV.

Strengthening the front armor along with installing new weapons resulted in making the vehicles nose-heavy and caused much wear of front-end parts in the bad road conditions in the East. This problem also appeared in North Africa, where the 5th Light (later 21st Armored) Division and the 15th Armored Division saw service as of 1941.

A thorough evaluation and a look at future development were given in the Reich Chancellery on November 29, 1941, on the subject of "Tank Production and Tank Defense.

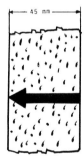

The penetration path of armor-piercing shells depending on the inclination of the armor plate.
1. Armor thickness 45 mm, 90-degree angle: penetration path 45 mm.

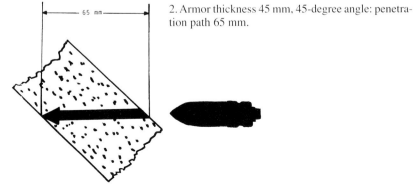

2. Armor thickness 45 mm, 45-degree angle: penetration path 65 mm.

A column of the 11th Armored Division on the road in the direction of Ostrog. In the first few weeks of the Russian campaign, remarkably long distances were marched. Correspondingly, technical breakdowns of armored vehicles were partially higher than the losses from enemy action.

A battery of Russian division artillery, armed with 76.2 mm Field Cannon 36 (F-22), has been overrun by German tanks. These guns, plentiful in the Red Army, could shoot down Type E Panzerkampfwagen III (as seen in the background) already at a range of 2,000 meters. To cover this distance, the tank needed five minutes, in which time the cannon could fire 40 to 50 shots.

In June 1941, at the beginning of the Russian campaign, the involved armored divisions had 763 Panzerkampfwagen 38(t) on hand. They saw action with the 7th, 8th, 12th, 19th, and 20th Armored Divisions.

Hitler stated then: "The experience gained in the eastern campaign proves that we stand at a turning point. Our antitank weapons are no longer equal to a part of the Russian tanks and the British infantry tanks. The armor plate of our tanks is no longer sufficient against the Russian antitank weapons." He said further: "Since poorly armored tanks are unusable, strengthening the armor must be undertaken, even at the cost of speed." As an important requirement, new designs were supposed to be limited to just three basic types.

1. A light type for use as a reconnaissance tank for the previous Panzerkampfwagen III.
2. A medium type for the formerly common Panzerkampfwagen IV.
3. A heavy type: a design by Henschel in collaboration with the Army Weapons Office and Dr. Porsche.

To conclude the discussion at the Reich Chancellery, Colonel Fichtner of Weapons Proving Department VI of the Army Weapons Office represented the views of the troops. They wanted a tank with long-range guns. Those were more important to them than heavier armor.

The defeats on the eastern front in the winter of 1941-42, with the resulting loss of large land areas, resulted in further extensive breakdowns of tanks and, what was much more serious, a shortage of battle-tested crews. In all, 2,758 tanks were lost in 1941, including 781 Type III and 369 Type IV.

At the end of 1941 a second series of Panzerkampfwagen III tanks, Type J, were delivered. This series was finally armed with the 5 cm Tank Gun 39 L/60. With the Antitank Shell 39, the result was a penetrating power of 72 mm at 500 meters; with the Antitank Shell 40/1 (hard-core), 76 mm of armor plate could be pierced at the same range.

The introduction of Panzerkampfwagen IV, Type F 2, with its longer 7.5 cm Tank Gun 40 L/43, later L/ 48, began in March 1942. At a range of 500 meters it was now possible to pierce 91 mm of armor with the Antitank Shell 39 and 108 mm with the (hard-core) Aintitank Shell 40.

This Panzerkampfwagen III (5 cm)(Sd.Kfz, 141), Type G, has sustained serious shot damage. Along with the T-34, which the Panzerkampfwagen III with its 5 cm L/42 tank gun could only deal with by hitting it in the sides and rear, the excellently armed Russian division artillery made its presence felt unpleasantly. The 76.2 mm Division Cannon 36 (F-22) could pierce 75 mm of armor plate at 500 meters (the Panzerkampfwagen III's armor in 1941 was 30 mm thick).

The German armored troops went into battle in 1942 with this tank. Front reports showed thoroughly positive experience. A report of July 9, 1942, on eastern experiences with tanks and antitank shells states: "...with the new long 5 cm and 7.5 cm tank guns, the troops have weapons in hand that have proved themselves splendidly in all battles. British tanks can be handled with assurance, and Russian tanks, except the strengthened KW-1, effectively with these weapons."

Experience gained in the combat in North Africa, such as a report of October 13, 1942, also stated that the Panzerkampfwagen III and IV with the short tank guns no longer lived up to the demands on them. A few days later, on October 22, 1942, the Wehrmacht High Command called for the hastened introduction of Panzerkampfwagen IV with the long 7.5 cm tank gun. Even the 5 cm antitank gun had remained ineffective against the new tanks used by the British.

But this rearming brought along a series of problems. In a report on the readiness for action of Tank Regiment 24 (24th Armored Division, ex 1st Cavalry Division) of March 21, 1942, we can read that 60% of the specified number of tanks was on hand, but the ammunition for the 5 cm Tank Gun 39 L/60, with which the division's 52 tanks were armed, was lacking.

Despite the better weapon effectiveness, there was still much emphasis on clever tactical maneuvering to gain an advantage in combat with enemy tanks. In a report on experience in fighting against tanks, dated October 17, 1942, it is said: "Also tanks with long tank guns must not move frontally against standing enemy tanks. Mobility and skillful utilization of the terrain assure superiority in a firefight." Tanks that were in action should be used effectively in mobile advances against the flanks and the rear of enemy armored units. Fulfilling all other tasks was given secondary importance.

In connection with the installation of long-barreled tank guns in Panzerkampfwagen III and IV and a general strengthening of the armor in the frontal areas of hull and turret, a noticeable decrease in tactical mobility of the vehicles had to be accepted. In Panzerkampfwagen III, 2,024 of which were on hand in April 1942, there was a considerable increase in the specific ground pressure (Kp per sq.cm.):

Panzer III, Type, Year	Fighting Weight	Specific Ground Pressure
A (1936)	18.3 tons	0.63 Kp/sq.cm.
B (1937)	18.5 tons	0.65 Kp/sq.cm.
E (1939)	19.5 tons	0.99 Kp/sq.cm.
H (1941)	21.6 tons	0.94 Kp/sq.cm. (wide tracks)
J (1941)	22.3 tons	0.95 Kp/sq.cm.
M (1942)	23.0 tons	1.05 Kp/sq.cm.

On May 10, 1940, the Army had 143 Panzerkampfwagen 35(t) on hand, 135 of which saw action in France. A good year later there were 187 of them on hand. Some 160 of them were used on the eastern front. In the picture, Tank Regiment 11 prepares to march in June 1941.

A total loss from hitting mines. Vehicles thus damaged were sent back to Germany. There the turrets were stored at army armories and later used in fixed fortifications. In March 1945 there were still 361 of them on hand.

Tank Regiment 21 had only 30 usable tanks left at the end of July 1941. Losses from enemy action and technical break-downs had considerably less-ened the fighting power of the 20th Armored Division.

This Panzerkampfwagen 38(t) was completely burned out after being hit in the hull by an antitank shell (between the two pairs of road wheels). In 1941, 773 tanks of this type were lost.

Reparable tanks were usually used to make self-propelled gun mounts. Some, though, were rebuilt for special pur-poses by the troops them-selves. This includes the 38(t) recovery tank shown in this picture.

This had its results in the complex terrain conditions on the eastern front. Here, the T-34 (with its specific ground pressure of 0.66 Kp per sq.cm.) could show clear advantages. Many German reports agreed on the "astonishing off-road capability" of the T-34, which enabled the Russian troops again and again to surprise the German antitank defenses.

How diverse the problems were that occupied the armored troops, and especially the already overworked repair services, is shown in the report, quoted below, of December 4, 1942. According to it, an armored division had lost 30% of its tanks because mice had gnawed on the electric wiring.

The year of 1942 brought success to the German troops near Kharkov. In due course the southern sector of the eastern front reached the Volga, and German troops advanced into the Caucasus. Again success was attained with decisive participation from armored troops, and in fact, against Russian tanks that were strengthened in every way. Again the year ended with serious losses. At Stalingrad the 6th Army, with parts of the 4th Armored Army, was surrounded. In the Stalingrad pocket three armored divisions were lost.

Attempts at action by the Hoth Army Group failed, although in many special actions modern tanks and antitank weapons had been received. The Don Army Group, as reported on December 21, 1942, received 162 Panzerkampfwagen III and 40 Panzerkampfwagen IV. For their support, the first Army tank unit with the new Panzerkampfwagen VI "Tiger" (Sd.Kfz. 181) was also sent into action here later.

A Panzerkampfwagen III (5 cm)(Sd.Kfz. 141), Type G of the 14th Armored Division at the eastern front in July 1941. Various reports are at hand on the practicality of the attachment of spare track links on the particularly exposed surfaces, which was increased at that time. The fact was that they offered no increase in protection from hits by large-caliber antitank shells.

In Armored Group 4, which belonged to the Army Group North, three armored divisions and two motorized infantry divisions saw service. They were the 1st, 6th, and 8th Armored Divisions, which attacked there on June 22, 1941, with something more than 570 tanks. The picture shows Panzerkampfwagen III (5 cm)(Sd.Kfz. 141), Type J, near Luga in August 1941.

A typical picture of the winter war of 1941-42 on the eastern front. Tanks and riflemen, accompanied by icy winds, laboriously move forward on the vast snowy plains. The penetrating power of tank wedges, which had advanced thousands of kilometers westward in the summer, was gone now.

A Panzerkampfwagen III (5 cm) (Sd/Kfz. 141), Type J is seen after being hit by heavy artillery fire. In 1941, 782 Panzerkampwagen III tanks were lost.

After the snow melted in the Kalinin area, the 1st Armored Division had to retreat after the attack on Moscow failed. Many damaged tanks were lost on account of bad weather conditions and the lack of means of recovery. For these reasons, the total losses in the armored units increased quickly during the withdrawal.

This Flamethrowing Tank II (Sd.Kfz. 122) was captured by the Russians before Moscow. On June 1, 1941, the Army had 85 vehicles of this type. They were created by rebuilding the Schnellkampfpanzer II, Type D-E.

On the eastern front in March 1942, a Panzerkampfwagen III of the 14th Armored Division is seen on its way from the repair shop to the fighting troops. In deep snow the comparatively heavy specific ground pressure of the German tanks soon became an obvious disadvantage as tanks got stuck in the snow.

Hitler described the Panzer-kampfwagen III, in a talk on November 29, 1941, as an "unsuccessful design." This evaluation is not justified. In its time this tank was a modern vehicle based on a progressive concept. Of course, when it encountered new generation tanks (T-34), it was inevitably inferior in a tactical-technical sense.

German tank losses in the winter of 1941-42 were large. The 7th Armored Division reported on January 27, 1942, that it had only five usable tanks. The situation was similar in other units.

Tanks sent to repair shops for major work were usually rearmed with long-barreled tank guns. Here, such a modernized turret is being prepared for mounting on the hull. The 5 cm L/42 tank gun weighed 400 kilograms, and the gun with the L/60 barrel weighed 435 kg.

Between March 1941 and July 1942, 1,549 Panzerkampfwagen III (5 cm)(Sd.Kfz. 141), Type J tanks were built, still armed with the short 5 cm tank gun. By December 1941 the first tanks with the long gun (39 L/60) were delivered. There were 1,067 of them in all.

A Panzerkampfwagen III (5 cm)(Sd.Kfz. 141), Type F stops at an obstacle in northwestern France in the summer of 1942. The tank bore added armor 30 cm thick with cutouts for the driver's visor and the Ball Mantlet 30.

In North Africa, technical breakdowns of the Panzerkampfwagen III tanks in use there were especially high. The dust-saturated air sucked directly into the engines eroded the cylinder walls, making them completely unusable.

In the North African stony wastes, the I./Tank Regiment 4 reported having 32 tanks on April 4, 1941. Six of them were Panzerkampfwagen I, twelve Panzer II, ten Panzer III, and three Panzer IV, plus one armored command car.

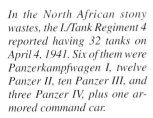

The Armored Command Car III (Sd.Kfz. 266, 267, 268), Type H, was built in a series of 145 vehicles as of November 1940, and 30 more were produced beginning in December 1941. The previous Types D and E, with quantities of 30 and 45 vehicles, came nowhere near the quantities of Type H.

This Armored Command Car III has a dummy gun that looks like the 5 cm L/42 tank gun which meanwhile had become the primary weapon of Panzerkampfwagen III. As of August 1942, armored command cars were also armed with this weapon.

In the summer of 1942 in the southern sector of the eastern front, tanks of the 24th Armored Division prepare for an attack. Panzerkampfwagen II and III were to be supported by Type IV tanks with long 7.5 cm tank guns in attacks on tank targets.

On the treeless steppes, the armored group of the 24th Armored Division, consisting of the tank regiment's vehicles and the armored personnel carriers of the mobile armored grenadier battalion (gp.), move eastward in August 1942. Their target was Stalingrad.

Tanks of the 24th Armored Division are seen on the Dun steppes in the summer of 1942. The armored command car is just crossing an antitank ditch.

Tanks and armored personnel carriers of the 24th Armored Division rest after breaking through the Russian defenses. The armored grenadiers have dismounted to round up Russian soldiers, whose stubborn defense always endangered the unarmored parts of the division and its urgently needed supply trains.

The second series of Type J Panzerkampfwagen III came armed with the 5 cm Tank Gun 39 L/60. Its shells reached an initial velocity of 835 m/sec, and the Tank Shell 40 even attained 1,180 m/sec. Compared to the L/42 tank gun, a penetrating power from 47 mm (55 mm) to 59 mm (72 mm) at a range of 500 meters was attained.

German tanks are seen on the Vanebiere of Marseilles in mid-November of 1942. The armored troops took part in the occupation of southern France, which had the code name "Anton." In Toulon there was even a fight between German tanks and ships of the French fleet.

Between October 1939 and May 1941, 229 Panzerkampfwagen IV (7.5 cm)(Sd.Kfz. 161), Type D, were built. The slightly strengthened armor offered no protection from Russian tank and antitank guns. This picture shows a fully burned-out tank of this type on a Russian road.

Production of Panzerkampfwagen IV (7.5 cm)(Sd.Kfz. 161), Type F, began in April 1941. One of its important improvements was the strengthening of the hull and turret front armor to 50 mm. The firms of VOMAG in Plauen and Nibelungenwerke in Sankt Valentin were engaged in building these tanks.

There were 462 Type F tanks built, most of them still with the short 7.5 cm tank gun. Later 25 tanks were armed with the new 7.5 cm Tank Gun 40 L/43.

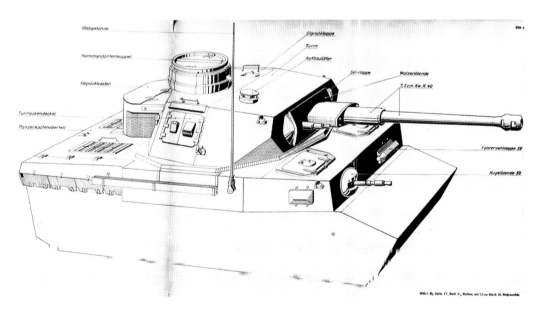

In a report of the 22nd Armored Division on May 20, 1942, about the use of armored vehicles with new weapons it was stated: "Maximum range at which the Kw K40 can fire effectively on tanks, some 1,200 meters." Direct hits meant the destruction of the tank, for which the Tank Shell 39 was fully sufficient.

Arming the Panzerkampfwagen IV with the capable 7.5 cm Tank Gun 40 L/43, first produced in March 1942, was urgently advocated at that time. Otherwise, the possibility of successful action against enemy tanks by the armored troops would be much less. In the East the armored divisions in Army Group South were first equipped with the new tanks.

The vehicles of three armored and four motorized rifle divisions were lost in the Stalingrad pocket. Russian collecting commands counted more than 500 tanks and assault guns, 130 of which were still usable at the end of December 1942.

1943:
Armored Troops in Times of Change

Generaloberst Guderian, in his memoirs, mentions a suggestion from front officers in 1941 that the Russian T 34 tanks ought to be copied. The reasons why this suggestion found no acceptance can surely be sought not in the manufacturing realm alone. Independently of the decision in this question, the wishes of the troops still showed very clearly that the German armored troops, if one wanted them to keep their tasks at a high level, had to be equipped with a new generation of tanks. They also made clear that all efforts to achieve increased fighting value through means of modernization had no perspective. After four years of service with the troops, the Panzerkampfwagen III and IV ranked as "scrap metal." Under wartime conditions the material and moral wear and tear were considerably greater. That caused Hitler, at the aforementioned meeting on November 29, 1941, to declare: "The age of the tank can be over soon."

Development contracts for the new generation of tanks had been given to the industry by November 1941. The future workhorses of armored warfare were to arise from two test vehicles, VK 3002 (later the Panzerkampfwagen V "Panther") and VK 3601 or VK 3002 (later Panzerkampfwagen VI "Tiger"). Without wanting to go into detail about a hectic developmental phase, marked by defeats and beset with problems, it can be said that they succeeded in getting the new tanks into existence within a year. In the midst of the war, that was a noteworthy achievement. In August 1942, the first eight Panzerkampfwagen VI "Tiger" E tanks (Sd.Kfz. 181) were delivered. The first two Panzerkampfwagen V "Panther" (Sd.Kfz. 171) followed in January 1943.

The short developmental time naturally had its price. The "Tiger" was in many ways a compromise, especially in terms of its angular armor plate. In the "Panther," with a much more modern design in this respect, faults in the power train became apparent. Progress was made in the area of armor plate (front armor 100 and 80 mm, respectively), whereby the laborious serration of the armor plates gave the armor greater rigidity. The 76.2 mm tank gun of the T-34 tank could no longer pierce the front armor of the new tanks. The same was true of the American and British tanks. On the other hand, the superior penetrating power of the new 8.8 cm Tank Gun 36 L/56 and the 7.5 cm Tank Gun 42 L/70 allowed the successful attacking of enemy tanks at long ranges. That this was attained with unaccustomed barrel lengths of 4.93 and 5.25 meters and could be disadvantageous in some combat situations—this consideration had at first prevented the introduction of a long 5 cm gun in 1940-41—was now a secondary matter.

The 7.5 cm Tank Gun 42 L/70, firing the Panzer Shell 39/42, could penetrate 99 mm of armor plate at a range of 1,500 meters (at a 60-degree angle). For the 8.8 cm gun, the figure at the same range was 91 mm. American tank commanders were astonished to find in Tunisia that their M4 Sherman tanks could be shot down by the "Tiger" at ranges of more than 2,000 meters. Since the beginning of 1943, Tank unit 501, one of the first Army units to be supplied with Panzerkampfwagen III, was in action there.

Along with the technical innovations tackled in building the tank, it was necessary to deal in a new way with all other questions relating to what was seen as a war-deciding weapon. The increasing superiority of the Russian armored troops forced this development. Generaloberst Guderian, transferred in December 1941 to the command reserve of the Army High Command, was regarded as the right man to solve the problems. Guderian requested extensive authority to fulfill his task. This was granted him in the "Service Instructions for the Inspector General of the Armored Troops" issued by Hitler on February 28, 1943. Structure and organization, training, and action of the armored troops were under the direction of the Inspector General of the Armored Troops. In addition, requests for further technical development of the tank weapon and for planning its manufacture were decided on in close collaboration with the Reich Minister for Armament and Ammunition.

Structure of the 7th Armored Division January 1943

(simplified portrayal)

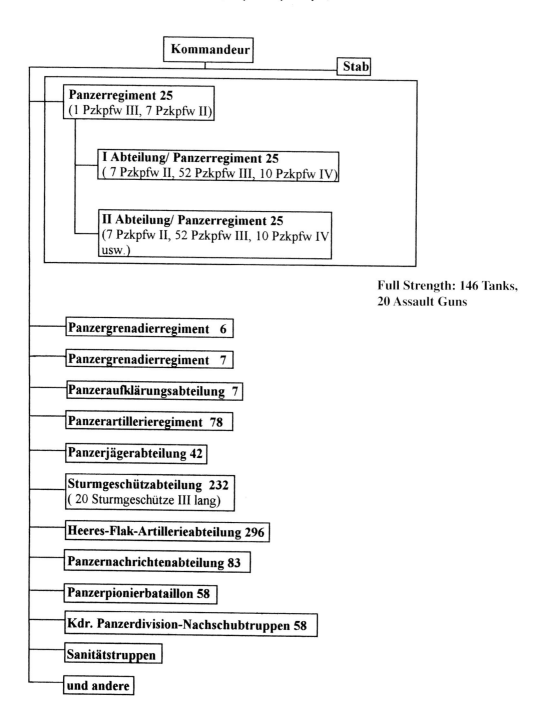

Kommandeur

Stab

Panzerregiment 25
(1 Pzkpfw III, 7 Pzkpfw II)

I Abteilung/ Panzerregiment 25
(7 Pzkpfw II, 52 Pzkpfw III, 10 Pzkpfw IV)

II Abteilung/ Panzerregiment 25
(7 Pzkpfw II, 52 Pzkpfw III, 10 Pzkpfw IV usw.)

**Full Strength: 146 Tanks,
20 Assault Guns**

Panzergrenadierregiment 6

Panzergrenadierregiment 7

Panzeraufklärungsabteilung 7

Panzerartillerieregiment 78

Panzerjägerabteilung 42

Sturmgeschützabteilung 232
(20 Sturmgeschütze III lang)

Heeres-Flak-Artillerieabteilung 296

Panzernachrichtenabteilung 83

Panzerpionierbataillon 58

Kdr. Panzerdivision-Nachschubtruppen 58

Sanitätstruppen

und andere

United under the designation of armored troops there were now also the armored grenadiers, infantry (motorized), armored reconnaissance troops, Panzerjäger, and the heavy assault guns. On March 9, 1943, Guderian outlined the tasks of the armored troops for the current year: A certain number of armored divisions should be available, fully combat-ready, for attacks on limited targets. For 1944, attacks in the grand style were planned once again.

The main driving force of the tank units in 1943 was the Panzerkampfwagen IV with frontal armor and track aprons. A limited number of tank units with the new Panzerkampfwagen V "Panther" and VI "Tiger" was expected, but their readiness for action was not to be reckoned on before July-August 1943. According to the construction program set up in September 1942, monthly production figures of 600 V and 50 VI were to be attained by the spring of 1944. In addition, 150 of the light reconnaissance tank ("Leopard") had been foreseen. In June 1943, 172 "Panther" tanks were delivered, followed by 213 in July. At the same time, 60 and then 65 "Tiger" tanks were delivered.

In order to get the armored divisions fully prepared for action, plans were made to equip one of the two tank units with assault guns as a transitional measure. Their monthly production at that time was around 300 vehicles.

For 1944, Guderian called for armored divisions with four units, which were to have a total of 400 tanks. That remained an illusion. The organization of "Armored Division 43," ordered for September 24, 1943, amounted to only one regiment of two units, with a total of 172 tanks. Units with the Panzerkampfwagen IV were to be given companies of 22 vehicles, units with the Panzerkampfwagen V "Panther" companies of 17 tanks each. Armored grenadier divisions were given an armored unit, which usually had three companies with 14 assault guns each.

The 7th Armored Division had two units of three companies each with 17 Panzerkampfwagen III and one company each with ten Panzerkampfwagen IV in its tank regiment on January 17, 1943. These Panzerkampfwagen III (5 cm L/60)(Sd.Kfz. 141/1), Type L tanks already had the newly introduced wider tracks with additional aprons.

In May 1941 the Henschel firm had received from the Army Weapons Office a contract to build a tank with 100 mm frontal armor, designated VK 3601. Its weight, originally set at 36 tons, went up to 55 tons from added requirements as a result of combat contact with the Russian T-34 and KW tanks.

The Army's tank units were regarded as especially important for the year 1943. At first they received a mixed supply of Panzerkampfwagen III and VI "Tiger" tanks, with units of 45 tanks each foreseen.

The intended situation as to the armored divisions for Operation "Citadel," the attack against the Russian front lines between Kharkov and Orel, still called for Panzerkampfwagen III with the long 5 cm and short 7.5 cm tank guns. In addition, the strengthened Panzerkampfwagen IV (Sd.Kfz. 161/2) with the 7.5 cm Tank Gun 40 L/48 was to be used. Even the Panzerkampfwagen II was still represented. It had already been certified in 1942 that in combat with comparable Russian tanks they were hopelessly inferior.

Typical of the uneven distribution of vehicles in the tank units are the SS Armored Grenadier Division "Das Reich" and the 7th Armored Division, both of which took part in Operation "Citadel."

The contracts for the development of the tank later known as the "Panther" had been issued on November 25, 1941. Series production began a good year later.

Vehicle	SS Panzergrenadier Division "Das Reich" (as of 7/2/1943)	7th Armored Division (as of 7/11/1943)
Panzerkampfwagen II (Sd.Kfz. 121)	—	4
Panzerkampfwagen III (Sd.Kfz. 141) with 5 cm KWK L/42	1	
Panzerkampfwagen III (Sd.Kfz. 141/1) with 5 cm KWK L/60	47	19
Panzerkampfwagen III (Sd/Kfz. 141/2) with 7.5 cm KWK L/24	—	5
Panzerkampfwagen IV (Sd.Kfz. 161) with 7.5 cm KWK L/24	—	1
Panzerkampfwagen IV (Sd.Kfz. 161/2) with 7.5 cm KWK L/48	29	25
Panzerkampfwagen VI (Sd.Kfz. 181) with 8.8 cm KWK L/56	12	—
Panzerbefehlswagen	8	4
Captured Tank T-34 with 7.62 cm KWK L/41.5	19	—
Total	116	53

Thus, up to seven different tank guns were used in a few divisions, and the required ammunition naturally had to be supplied for them—a problem which did not exactly make supplying easier.

The Inspector General of the Armored Troops tried harder and harder to correct the numerous faults and weak points in the action of the tank groups. For example, in a memo of April 27, 1943, he criticized the fact that tank crews were abandoning their tanks after minor damage. The widespread divided actions of tank companies and platoons should be a thing of the past. Tank officers described such actions as "publicity stunts," and spoke against the misuse of their tanks as "moral corset stays of the infantry." Tanks were lost only because means of recovery were lacking more often than was the case with complete units. The same is true of the use of tanks, often practiced in the past, in city street fighting. On October 11, 1942, after several months of bitter street fighting, the 24th Armored Division had summed up: "The appearance of terrain difficulties, such as house ruins, bomb craters, minefields, tank traps, and road blocks limited the mobility and observation ability of the tanks considerably, so that the use of tank units in street fighting is to be rejected on principle. The resulting losses are not justified by the successes of the Army's most valuable weapon. The main weapons of the tank, 'fire and movement,' do not have any effect...."

On the night of July 3-4, the last parts of the attack divisions intended for "Citadel" took up their readiness positions around the Kursk Bend. On the morning of July 5 the German attack began. North and south of Kursk two battles raged, in which the Germans were using, among others, 17 armored divisions, two armored grenadier divisions, and nine assault gun units. In addition to 10,000 guns and grenade launchers, 2,700 tanks and assault guns (46% of the entire supply) saw service. The Red Army troops were numerically superior in all positions on the battlefield; they could bring 3,300 tanks and self-propelled guns into action. That was 35% of their entire supply—which shows the quantitative superiority meanwhile attained by the Russian side.

The enemy tank of 1943: A MK IV "Churchill" IV infantry tank, delivered to the Soviet Union as part of the lend-lease agreement. It also saw service in Tunisia, Italy, and north-western Europe.

Several variations of the Russian KW tank were produced. At the beginning of July 1943, the Red Army, according to its own statistics, had 9,580 tanks, of which 6,232 were heavy or medium. This KW-1 was shot down in the Kursk area in July.

More than 35,000 of the American "Sherman" tanks were built for the war. The M 4 A 1 version alone was produced in a series of 6,281 tanks.

In the German attack group there were 200 "Panther" and 129 "Tiger" tanks, plus 90 heavy "Ferdinand" tank destroyers. The Germans' great hopes for them were only partly realized. Of the 200 "Panther" tanks in Tank Brigades 51 and 52, serving in the sector of the "Grossdeutschland" Armored Grenadier Division, only 40 were still ready for action at the end of the first day of attacking. 36 Panzerkampfwagen V "Panther" tanks had to be written off as total losses, while the other losses were attributable to technical faults.

In the course of combat at Kursk, the German armored troops applied a new attacking process, called the "Tank Bell" (Panzerglocke). In it, heavy Panzerkampfwagen VI "Tiger" tanks were in the middle of the attack formation. In an arc extending to the back and both sides, Panzerkampfwagen III and IV followed. Behind the center, light tanks were held ready in case of a breakthrough into the enemy defenses. In opposition to the previously usual "steel core" tactic, in which the more heavily armed Panzerkampfwagen IV, as surveillance tanks, primarily were to fight the targets more capable of resistance. Here it was a matter of being able to conduct the firefight across the entire width of the attack front. Using the new and potent tank guns made it possible. On the other side, the increasing effectiveness of the Russian tanks and antitank defenses demanded such an attack procedure.

The high point of the heavy fighting around the Kursk bend was doubtless the tank battle near Prokorovka on July 12, 1943. The German forces strove for an operative breakthrough here with about 500 tanks and assault guns. The Russian side did everything to prevent this and sent the 5th Guard Tank Army with some 850 tanks and self-propelled guns into the battle.

A Heavy Charge Carrier B-IV (Sd.Kfz, 301)m Type B, of Tank Unit (FKL) 301 in the northern sector of the Kursk Bend in July 1943. Production of this type of vehicle was halted in September 1944. In order to handle the strengthened enemy antitank defense, especially the use of mines, Guderian already advocated the monthly production of 150 charge carriers in December 1943, though a 60% lighter type, the NSU "Springer" (Sd.Kfz. 304).

In front of the driver on this Panzerkampfwagen III (7.5 cm)(Sd.Kfz. 141/2), Type N, the troop symbol for Radio-link Tank Group 301 can be seen. The photo was taken in the Kursk area in July 1943.

For special purposes the troops kept older tanks in action almost unchanged. Here a Panzerkampfwagen III (5 cm)(Sd.Kfz. 141), Type J, serves as a medical tank in the northern sector of the Kursk Bend. At left is a Panzerkampfwagen III (7.5 cm)(Sd.Kfz. 141/2), Type N. Heavy "Ferdinand" tank destroyers can also be seen.

On a section of terrain some six kilometers wide the largest tank battle took place. An eyewitness of this battle, Marshal Vassilevski, reported in Moscow on July 13, 1943: "Yesterday I personally observed a tank battle of our 18th and 29th Corps and over 300 enemy tanks in a counterattack southwest of Prokorovka. Simultaneously, hundreds of guns and all of our available launchers took part. After one hour, the whole battlefield was littered with burning German and Soviet tanks."

The German attack, after heavy fighting, lasted until the evening of July 12 on the north shore of the Psjol, where it finally came to a stop. The tank units were exhausted; reserves were scarcely to be had. After a conference on the next day at the Führer's headquarters, the offensive was ordered broken off.

It is very rewarding to study the loss figures reported by both sides. On August 7, 1943, the Soviet Information Bureau reported the German tank losses for the July 7-August 6, 1943, period at 4,605 destroyed and 521 captured tanks. That would have been a total of 5,126 total losses, a figure that would have eliminated the German armored troops. Early in July, just before Operation "Citadel" began, the total number of tanks in the Wehrmacht was stated as 5,850 on all fronts. Here the Soviet Information Bureau has exaggerated extremely.

The unsuccessful July combat actually resulted in the loss of some 1,500 tanks and assault guns. These losses were an extremely hard blow for the armored troops, which were still being reorganized. It had been the armored divisions that bore the heaviest burden of the battles. In seven of the twenty armored groups that saw action, more than 50% of their personnel had been lost. Experienced cadres could not be replaced very quickly by the hastily trained replacements from Germany. The increasingly hard fights left less and less time to gather new experience.

Guderian evaluated the possibilities of the armored troops after Kursk as follows: "The armored forces refreshed with great trouble were unusable for a long time because of heavy losses in manpower and materials." In his memo of October 17, 1943, the Inspector General of the Armored Troops even drew conclusions like these: "Waging war in the immediate future will not at first allow large-scale offensive operations....," and "If the tank weapon was hitherto one of the means of deciding the war, now it decides the outcome of the war itself."

In November 1942 the Army General Staff estimated that the Panzerkampfwagen III with the 7.5 cm Tank Gun 37 L/24 is considerably more effective than with the earlier 5 cm gun. Since the barrel was also used for the 7.5 cm assault gun on the Medium Infantry Tank (Sd.Kfz. 251/9), the request for renewed production of this weapon was approved.

The Armored Observation Vehicle III (Sd.Kfz. 143) was built in a series of 262 vehicles from February 1943 to April 1944. In the units of the Armored Artillery regiments equipped with "Hummel" and "Wespe" self-propelled guns, every battery had such vehicles since 1943, as well as a fire-control Schützenpanzerwagen.

The setup of tank weapons for attacking armored targets that took place during the war increased the wish for armored vehicles whose armaments could achieve a great explosive and splintering effect. In Germany this produced the Assault Tank IV (Sd.Kfz. 166) "Brummbär" (15 cm Assault Howitzer 43). 306 of these vehicles were built.

Originally, the "Ferdinand" bore the designation "Assault Gun with 8,8 cm Pak 43/2 (Sd.Kfz. 191)." In July 1943, two Army Panzerjäger units, each with 45 of these 65-ton monsters, saw service in the northern sector of the Kursk Bend. The last vehicles were lost at the end of April 1945 near Gross-Köris, south of Berlin.

Let us turn now to the Red Army's tank losses. In an outline of the Foreign Armies East Department (II d) about the armament industry and tank production in Soviet Russia, the Russian side's losses of tanks and self-propelled guns in July 1943 are given at 4,150. The troops had reported over 8,000, but 50% of them were subtracted as double reports. For the same month, the production of 2,160 tanks and self-propelled guns was reported. In addition, there were Lend-Lease deliveries from the USA and Britain. According to J. Magnuski, there were 24,188 tanks and self-propelled guns built in 1943. The Soviet Union, accordingly, needed two months' production to make up for the heavy losses of July 1943. According to German information, the number of tanks and self-propelled guns in the Red Army in August 1943 was 12,660 vehicles under the July total, but was still more than double the number of the Wehrmacht's vehicles. In all, there were about 5,500 vehicles available on all fronts; monthly production reached 1,658 tanks and assault guns in July. The yearly production could be raised to 12,075 vehicles. In spite of that, the tank war had been decided. Germany had lost it on the assembly lines.

The effects were clear to see. Since the manufacture of spare parts had been cut back for the sake of production of complete armored vehicles, parts were lacking everywhere. This is seen in a message from Army Group South on September 17, 1943: "The shortage of tank motors now affects our fighting power considerably.... For example, the SS Division 'Das Reich' has reported that the tank regiment needs motors for ten Panzerkampfwagen IV, 20 Panzerkampfwagen V, two Panzerkampfwagen VI, and seven Panzerkampfwagen III (Assault Gun III), and that the regiment's readiness for combat is only guaranteed for a few days. In addition, the other seven armored divisions need motors for the following armored vehicles:

- 28 Panzerkampfwagen III
- 46 Panzerkampfwagen IV
- 10 Panzerkampfwagen V
- 5 Panzerkampfawgen VI

Fast supplying of at least a portion of the motors by air transport is urgently requested."

Along with the notably more effective explosive shell, the 7.5 cm tank gun could fire Grenade Cartridge 38 HL/ A, B, or C. Depending on which type was used, they could pierce 70 to 100 mm of armor plate at ranges up to 1,500 meters. Compared to the previous armament of the Panzerkampfwagen III with the 5 cm tank gun, performance was increased threefold.

Panzerkampfwagen III and IV of Tank Regiment 3 of the 2nd Armored Division are seen after the end of Operation "Citadel." Under this code name the German attacks in the Kursk Bend were carried out in July 1943.

Early in March 1943, firing tests on tank aprons for Panzerkampfwagen III and IV were concluded successfully. Thereupon Hitler ordered the fitting of such aprons on all newly-built tanks. Vehicles already with the troops were to be re-equipped.

As can be seen from the side hatch in the hull of this tank, it is an older-type Panzer III. With added armor in the driver's area and a 7.5 cm gun, it was brought up to the Type N norm.

Between June 1942 and August 1943, 663 Panzerkampfwagen III (7.5 cm)(Sd/Kfz. 141/2), Type N, were built. There were also 37 made by rebuilding other vehicles.

Panzerkampfwagen III with the 7.5 cm tank gun were not found only in the tank units of the armored and armored grenadier divisions. The Army's first units equipped with "Tiger," tanks and the radio units were also equipped with them.

A Panzerkampfwagen III (5 cm L/60)(Sd.Kfz. 141/1), Type M (note the raised muffler) of the 2nd SS Armored Division "Das Reich" is seen at the western edge of Kharkov on March 11, 1943, as the city was recaptured.

A Panzerkampfwagen III (5 cm L/60)(Sd.Kfz. 141/1), Type L secures the tank repair shop of Tank Regiment 33 against breakthroughs by Russian tanks in February 1943.

Front view of a Panzerkampf-wagen III, Type L (18th Armored Division), showing the 20 mm frontal armor in the driver's area.

German tanks in Finnish woodlands. The sending of individual tank companies was to be evaluated in reference to Germany's allies. In this picture is a Panzerkampfwagen IV (75.5 cm L/48)(Sd.Kfz. 161/2), Type H, of which 3,774 were produced between April 1943 and July 1944.

In 1943 the Panzerkampfwagen IV formed the backbone of the tank regiments in the armored divisions. Gradually they replaced the Panzerkampfwagen III, so that at least one unit in each regiment could be supplied with 17 takns of this type. In that year 3,073 Panzerkampfwagen IV were built.

Fast changes in the weather forced the troops to improvise when camouflaging their vehicles. When snow fell, a chalky liquid was applied. Often the tanks were smeared with lime, and sometimes used oil with sand thrown on it.

In May 1942 the F-2 version of Panzerkampfwagen IV (7.5 cm L/48)(Sd.Kfz. 161/1) was followed by Type G, with the final form of the long 7.5 cm tank gun. By June 1943, 1,687 vehicles of this type had been delivered. This picture was taken on the Peloponessus in Greece on August 4, 1943.

A newly equipped tank unit is being transferred near Argos, Greece, on August 4, 1943.

This Panzerkampfwagen V (Sd.Kfz. 171) "Panther," Type A, already shows the new ball-shaped gun mantlet for the radioman's machine gun in its hull.

Between January and September 1943, 850 of the Panzerkampfwagen V (Sd.Kfz. 171) "Panther" Type D tank were built. They were very vulnerable mechanically, and did not make a good impression on the troops in their first use at the Kursk Bend.

Many Type D "Panthers" remained in Germany after being repaired, in order to be used in training, as was the one shown here. To save weight, the weapons and some other components were removed.

After a year's development, series production of the Panzerkampfwagen VI (Sd/Kfz. 181) "Tiger" I began in July 1942. By May 12, 1943, the first 285 vehicles had been delivered.

The "Tiger" tank's first action suffered from their short developmental time, insufficient testing, and the rushed formation of the first units, with numerous shortages. Here, three 18-ton tractors (Sd.Kfz. 9) tow a damaged "Tiger" to the repair shop. The picture was taken early in March 1943 west of Kharkov.

A "Tiger" of Heavy Tank Unit 503 (a tiger's head was the troop's emblem) at Mariupol in April 1943. This unit also had a mixed supply, and only gained a third company in February of that year.

Otto Carius, Oberleutnant in Heavy Tank Unit 502 during the war, wrote in his book "Tigers in the Mud": "Again and again we admired the quality of the steel in our tanks. It was hard without being brittle, and also elastic despite its hardness...."

The "Tiger" was used by the heavy tank units to form focal points in armored and armored grenadier divisions. Its armament and heavy armor, in particular, let the tanks make decisive advances. It was also intended for this use in the Kursk Bend combat, where more than 130 "Tigers" saw action.

According to a memo of January 2, 1943, 30 "Tiger" tanks were taken in December 1942. The production of tank guns had to be cut back from 25 to 16 in January 1943, in favor of the 8.8 cm Assault Gun 43 for the "Tiger" (P) tank destroyer. Vehicles that were delivered were sent on to the troops as quickly as possible.

In May 1942, Heavy Tank Unit 501 was formed, and in that November it was sent to North Africa. In this theater of war the desert landscape offered a sufficiently wide range for the 8.8 cm Tank Gun 36 L/56, which could pierce 84 mm of armor plate even at a range of 2,000 meters.

Despite many weak points, which included the angular armor, Panzerkampfwagen VI (Sd./Kfz. 181) "Tiger" was a respected tank. Even at ranges under 100 meters it was not possible for the Russian T-34, with its 76.2 mm F-34 L/41 tank gun (84 mm penetration at 100 meters), to penetrate the "Tiger's" frontal armor.

Above: *In the summer of 1943 the "Tiger" unit gradually received its full quota of 45 Panzerkampfwagen VI (Sd.Kfz. 181) tanks. By August 1944, 1,354 tanks of this type had been delivered. This was not a large number; the Foreign Armies East Department reported that the Soviet Union had built 20,350 tanks in 1943. Below: In the morning hours of July 5, 1943, an avalanche of tanks rolled toward the Russian positions in the Kursk Bend. At the head of the formations were "Tiger" tanks and heavy "Ferdinand" assault guns. The combat around Kursk included the largest tank battle in history.*

Oberstleutnant Dr. Franz Bäke commanded the Bäke Heavy Tank Regiment early in 1944, including Heavy Tank unit 503 with 34 "Tiger" tanks. In a six-day tank battle near Balabanovka the group shot down 267 enemy tanks.

This tank has thrown a track; while the damage is being repaired, the turret crew watches for attacking Russian tanks. Often it was such minor damage that led to the total loss of a tank, because means of recovery were lacking, and the territory had to be given up.

1944:
Armored Troops on the Defensive

The course of battle until the war's end was foreshadowed clearly in 1943. The Russian armored troops, becoming more and more superior, became the deciding factor on the battlefield. General Buhle, Chief of Wehrmacht Armament, looked back at the beginning of 1945 and stated: "All battles in the recent past gave the clear picture that only the tank had been really decisive."

Germany was pushed into strategic defense. There the armored troops were the most important element in an active defense. Already in 1943, and much more so in 1944, the following variations of action for armored units under changed conditions emerged:

1. Counterstrikes or counterattacks with tank units in operative and operative-tactical order of extent.
2. Maneuvers with armored divisions as intervention reserves in the form of so-called "fireman" actions.
3. The actions of tank units in defending positions.

The dominant form of tank use became the counterattack.

The changed forms of tank use were linked with a series of problems. Splinter action led to unbearable burdens on the weakened supply services. In 1943 one armored division needed 600 tons of ammunition, 600 tons of supplies and 185 cubic meters of fuel per month. During the war, fuel supplying became a general problem. But there were also shortages in ammunition supply. In September 1943, the III. and XXXXVIII. Armored Corps complained about the tense ammunition situation for the 7.5 cm Tank Gun 40 L/43 and L/48. Only 16, or 1% of the specified supply of explosive shells, and 44, or 42% of antitank shells were available. Tank repairs were made difficult. The frequent maneuvering with the armored parts of the division raised the number of technical breakdowns. The repair units were often detached from the combat troops. Thus, the punctual recovery and repair of the vehicles was imperiled, and in numerous cases they became total losses when territory was given up, often because of minor damage. In October 1943 only about 300 of the 3,000 tanks on the eastern front were ready for action. The Inspector General of the Armored Troops tried through numerous basic commands, including that of December 7, 1943, to counteract this bad situation, but without success. Often sufficiently strong means of recovery for the ever-heavier tanks were lacking. Recovery tanks came into being through rebuilding, sometimes singly.

The purely technical development of German tank technology, especially noticeable in the realm of higher-performance weapons and improved armor protection, became part of the clearly changing manner of tank use that was seen in 1943. The development proceeded toward tanks of strongly defensive character. The already mentioned equipping of a number of tanks with assault guns must also be evaluated in this direction.

The Panzerkampfwagen V "Panther" (Sd.Kfz. 171) was described in a French evaluation report of 1947 as lacking strategic mobility. The reasons for this were the meager range and the mechanical vulnerability of the power train. This was naturally true of the Panzerkampfwagen VI "Tiger" (Sd.Kfz. 181), too. The fighting weight of these tanks, 46 and 55 tons, respectively, was too high. Moving tanks of this weight class to the scene of action over too-weak bridges and swampy ground, to name only two types of hindrances, often required all-out efforts from their crews and often led to the loss of the vehices. Successful actions, like those of SS Obersturmführer Wittmann (Heavy SS Tank Unit 101), who was able to shoot down a whole column of the 7th British Armored Division in Normandy on June 13, 1944, were rare and attributable to favorable circumstances. In such cases the superior weapons and armor of the "Tiger" could be utilized fully.

A Panzerkampfwagen III (5 cm L/60)(Sd.Kfz. 141/1), Type M, leads a column of assault guns to prepare for a counterattack, along with a "Tiger" company on the eastern front in the winter of 1943-44.

A further reason for the limited mobility of the German tank units was the loss of air superiority at the fronts. The Germans gained their first experience of this in North Africa in the spring of 1943. After the Normandy landing in June 1944, German tank units that were supposed to drive the landing troops back were regularly "nailed to the ground" by Allied fighter-bombers.

Vehicles of the "Panther" unit of SS Tank Regiment 5 (5th SS Armored Division "Viking") on the eastern front in the spring of 1944.

The continuing heavy demands on the armored divisions prevented the general refreshing of the units already planned by Guderian in 1943. According to the 1944 armament plan, the Army's organizational department reckoned on equipping each of the existing 32 armored divisions with 130 Panzerkampfwagen IV (Sd.Kfz. 161/2) and V "Panther" (Sd.Kfz. 171).

The number of heavy armored units in the Army equipped with Panzerkampfwagen VI "Tiger" tanks (Sd.Kfz. 181) was to be increased from 12 to 14. In addition, some tank units were supplied with "Panthers." A restructuring, coupled with the hope of improving the armored troops' many "fireman" actions, did not prove itself in the end.

At the beginning of 1944, the following checklist of equipment was made up for the armored troops:

Type of Weapon	Number on hand	Notes
Panzerkampfwagen III Flamethrower (Sd.Kfz. 141/3)	7 per platoon	one-time production, 1943
Armored Observation Car III (Sd.Kfz. 143)	4 per armored division	number needed
Panzerkampfwagen IV (Sd.Kfz. 161/2)	96 per unit 22 per Co. = 88 +8 in staffs	uniform equipment
Panzerkampfwagen 5 "Panther" (Sd.Kfz. 171)	76 per unit +2 recovery tanks, 17 per Co. = 68 + 8 in staffs	uniform equipment later 96 planned
Panzerkampfwagen VI "Tiger" (Sd.Kfz. 181)	45 per unit	
Heavy Charge Carrier B-IV (Sd.Kfz. 301)	36 per company	12 of them substitutes
Assault Gun 40 and 40 n.A. (Sd.Kfz. 142/1 and 163)	31 per unit/45 per unit 14 per Co. = 42 + 3 in staff, 10 per company	in artillery/armored troops ID Panzerjäger companies
Assault Howitzer 42 (Sd.Kfz. 142/2)	9 per unit	in the artillery then only 22 Assault Gun 40
Tank Destroyer "Ferdinand" (Sd.Kfz. 184 s)	45 per unit	
Heavy Tank Destroyer ("Panther")	45 per unit	
Superheavy Tank Destroyer ("Tiger")	45 per unit	
Assault Tank IV "Brummba"r") (Sd.Kfz. 166)	45 per unit	
Gun Wagon II "Wespe" (Sd.Kfz. 124)	6 per battery	
Gun Wagon III/IV "Hummel" (Sd.Kfz. 165)	6 per battery	

15 cm Heavy Infantry Gun 33 (Sf.)(Sd.Kfz. 138/1)	6 per infantry gun company	in armored grenadier regiment
Pak 40 (mot Z)	12 per Co./9 per Co.	Panzerjäger/Grenadier units
Pak 40 (Sf.II)(Sd.Kfz. 131)	14 per company	
Pak 40 (Sf. 38)(Sd.Kfz. 138)	14 per company	
Pak 43/41	36 per unit	of three companies
"Hornisse"	45 per unit	
Light SPW (Sd.Kfz. 250) & medium SPW (Sd.Kfz. 251)	89 med. SPW per armored grenadier battalion, also in armored engineer company, another 24 medium SPW, 89 light & 26 medium SPW per armored reconnaissance unit	
Medium flame SPW (Sd.Kfz. 251/6)	6 per platoon	
Light 4-whl. armored scout car (Sd.Kfz. 222)	14 per armored scout company	
Heavy 8-whl. armored scout car (Sd.Kfz. 231 & 234)	6 per armored scout company	
Armored Scout Car "Luchs" (Sd.Kfz. 123)	29 uniform per armored scout company	

The Assault Gun III (Sd.Kfz. 142/1), Type G, was among the Army's best-known armored vehicles. Some 7,800 of them were built up to March 1945. They were used particularly in the assault gun units, and also as transitional equipment in the tank-destroyer and tank units.

In December 1943 the General Staff considered cutting back production of the medium armored personnel carrier in favor of the 3-ton tractor. The Inspector General of the Armored Troops voiced his position: The maintenance of strong armored units is inseparably linked with the maintenance of the fighting power of the armored grenadier regiments. The battalions equipped with armored personnel carriers had considerably fewer losses than the unarmored grenadier battalions.

By March 1945 almost 500 "Nashorn" (Rhinoceros) tank destroyers (originally "Hornisse," or Hornet) had been built. This vehicle, armed with the powerful 8.8 cm Antitank Gun 43/1 L/71, was also used by Army tank-destroyer units.

In 1944 the tank-destroyer units of the armored divisions were assigned two companies, each with 14 Tank Destroyer IV (Sd.Kfz. 162/1), Type F. In February 1945 Tank Destroyer Unit 53 (5th Armored Division) reported: The Tank Destroyer IV, in technical terms, does not meet its requirements in any way. Of 14 total losses, only three were caused by enemy action; 11 tank destroyers had to be blown up for lack of towing tractors.

Structure of the 5th Armored Division 1944

(simplified portrayal)

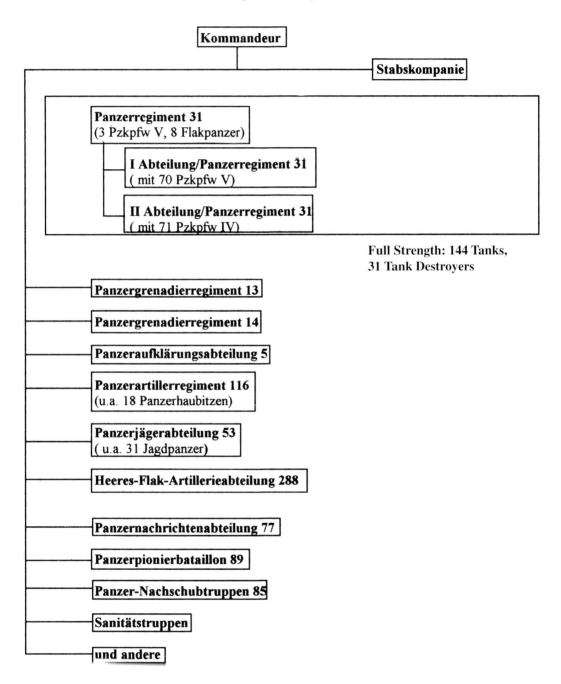

Kommandeur

Stabskompanie

Panzerrcgiment 31
(3 Pzkpfw V, 8 Flakpanzer)

I Abteilung/Panzerregiment 31
(mit 70 Pzkpfw V)

II Abteilung/Panzerregiment 31
(mit 71 Pzkpfw IV)

Full Strength: 144 Tanks,
31 Tank Destroyers

Panzergrenadierregiment 13

Panzergrenadierregiment 14

Panzeraufklärungsabteilung 5

Panzerartillerregiment 116
(u.a. 18 Panzerhaubitzen)

Panzerjägerabteilung 53
(u.a. 31 Jagdpanzer)

Heeres-Flak-Artillerieabteilung 288

Panzernachrichtenabteilung 77

Panzerpionierbataillon 89

Panzer-Nachschubtruppen 85

Sanitätstruppen

und andere

Older tanks, most of which came from repair shops, were assigned, until the war ended, to the armored replacement and training units. In this picture a Panzerkampfwagen III (5 cm)(Sd.Kfz. 141), Type G, with added armor is being used for training at the Ohrdruf training camp.

These Panzerkampfwagen III with box running gear and gasoline power were used by Tank Training Unit 10 in Eisenach, as seen in the spring of 1944.

Tank close-combat training with a Type N Panzerkampf-wagen III, photographed in the spring of 1944 during an officer candidate training course run by Tank Training Unit 10 in Eisenach.

Development of tank penetrating performance by the primary weapons of German tanks between 1935 and 1945

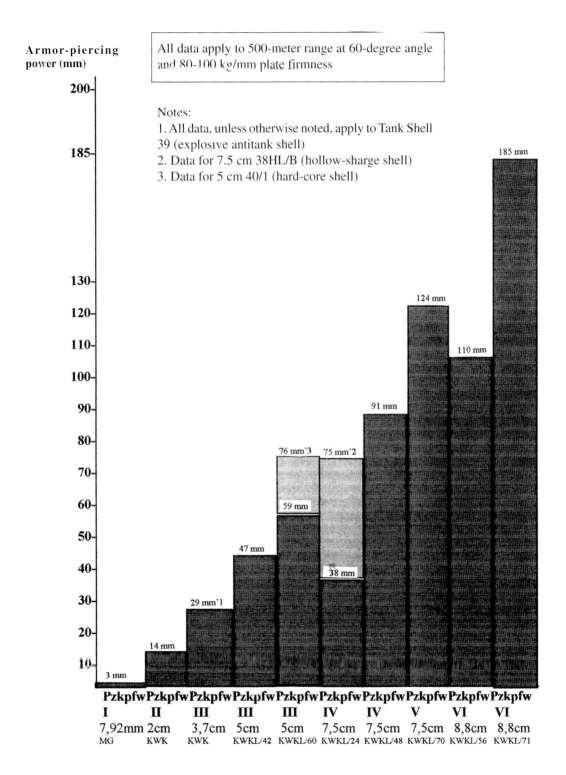

Armor-piercing power (mm)

All data apply to 500-meter range at 60-degree angle and 80-100 kg/mm plate firmness

Notes:
1. All data, unless otherwise noted, apply to Tank Shell 39 (explosive antitank shell)
2. Data for 7.5 cm 38HL/B (hollow-sharge shell)
3. Data for 5 cm 40/1 (hard-core shell)

3 mm	14 mm	29 mm'1	47 mm	59 mm	76 mm'3 / 38 mm / 75 mm'2	91 mm	124 mm	110 mm	185 mm
Pzkpfw	Pzkpfw	Pzkpfw	Pzkpfw	Pzkpfw	Pzkpfw	Pzkpfw	Pzkpfw	Pzkpfw	Pzkpfw
I	II	III	III	III	IV	IV	V	VI	VI
7,92mm	2cm	3,7cm	5cm	5cm	7,5cm	7,5cm	7,5cm	8,8cm	8,8cm
MG	KWK	KWK	KWKL/42	KWKL/60	KWKL/24	KWKL/48	KWKL/70	KWKL/56	KWKL/71

For the complete supply of an "Armored Division 44," varying figures were again valid. According to them, the tank regiment should have a total of 149 tanks; in each of the two tank units there were to be four companies uniformly equipped with 17 Panzerkampfwagen IV (Sd.Kfz. 161/2) and Panzerkampfwagen V "Panther" (Sd.Kfz. 171), a total of 136 vehicles. The remaining tanks were with the unit staffs or regimental staff. An important increase in the armored components of the divisions was the equipping of each of two companies of the tank-destroyer unit with 14 Tank Destroyer IV (Sd.Kfz. 162). There were also three of them in the unit staff.

A reduction of tank and armored vehicle types from 41 to 14 was planned, which was to be part of the standardization process slowly taking place in the industry. In addition, this should make it considerably easier to supply spare parts and make repairs among the troops. Intervention in the production of equipment could be made only with great care; otherwise, declining production figures would have a negative effect on the supplies of armored vehicles at the fronts.

By the end of 1944 it was possible to withdraw twelve armored and armored grenadier divisions from the front and refresh them according to the "Armored Division 44" structure. In practice, though, the dividing of armored units into battle groups took place. They formed the backbone of the forces kept ready for counterattacks. An independent tactic was developed. This is noted in an experience report of the 13th Armored Division of January 13, 1944, stating that armored battle groups were utilized as tactical intervention groups to reestablish the main battle line or connections with cut-off troop units, cutting off and making counterthrusts. As a rule, such battle groups acted from the second position. They combined high firepower, mobility, and armor protection and thus, thanks to the resulting fighting and thrusting strength, use local surprise attacks to compensate for the enemy's general superiority, at least temporarily. A typical example is the counterattacks of the IV. SS Armored Corps northeast of Warsaw, as a result of which the 2nd Armored Army on the 1st Belarus Front suffered a serious defeat early in August 1944. In 1944 it was customary to delegate armored battle groups to neighboring sectors to help master critical situations.

The armor plate that was 180 and 150 mm think on the front of the hull and turret gave the "Tiger" II reliable protection from enemy fire. Along with its strong weaponry, it won the respectful nickname of "King Tiger" from the Americans. To be sure, all this was gained for the price of the impractical weight of over seventy tons.

The Panzerkampfwagen IV (7.5 cm L/48)(Sd.Kfz. 161/2), Type J, built as of June 1944, had its front hull armor, like that of the Type H, increased to 80 mm. This picture was taken at a training unit's camp in Denmark.

Along with its excellent penetrating power against armor plate, the 7.5 cm Tank Gun 42 L/70 showed high targeting accuracy: At 500 meters all of ten shots were expected to hit the target; at 2,000 meters only two or three.

In the course of "Panther" production, its mechanical reliability could be improved through innovations that constantly were added. All the same, the railroad was used even for transfer over short distances. Naturally, the high fuel consumption, between 450 and 670 liters per 100 kilometers, played a major role here.

To produce one Panzerkampfwagen V (Sd.Kfz. 171) "Panther," the needed raw materials included 77.5 tons of iron (not counting weapons). The average production time was 14 months, and the price was about 117,000 Reichsmark.

In the 1944 armament study, the Inspector General of the Armored Troops requested the establishment and equipping of 34 armored divisions, requiring an average monthly production of 400 "Panther" tanks.

This rear view of a Type G "Panther" shows the sharply downward-sloping edge of the armored hull over the running gear. This new shape for the hull resulted in the production of uniform hulls for the "Panther" and "Jagdpanther."

A newly established "Panther" unit with vehicles of Types A and G prepares for an attack. The Panzerkampfwagen V (Sd.Kfz. 171) "Panther," Type G, was built as of March 1944. This picture was taken in Poland in the summer of 1944.

According to Russian evaluations, the "Tiger" was a powerful tank in terms of its arms and armor, although the armor did not have optimal angles. It was particularly noted that it was easy to drive despite its heavy weight.

Heavy Tank units 504 and 508 saw action in the Italian campaign. The unfavorable terrain and Allied air superiority had negative effects on the use of the ponderous "Tiger" tank there. (BA)

The heavy IS-2 tank was used as of 1944 in the heavy guard tank regiments of the Red Army. In November 1944, heavy guard tank brigades of three regiments, each with 65 "Stalin" tanks, were first formed. The Russian answer to the German "Tiger" weighed only 46 tons, had up to 160 mm of armor, and carried a 122 mm tank gun.

A display of armor-piercing ammunition for the primary weapons of German tanks between 1935 and 1945 (from right to left): 1. 7.92 mm S m.K.L. 'Spur bullet (MG 13); 2. 2 cm tank shell L'Spur (2-cm KWK L/55); 3. 3.7 cm tank shell (3.7-cm-KWK L/45); 4. 5 cm tank shell 39 (5-cm-KWK L/42); 5. 5 cm tank shell 40/1 (5 cm-KWK L/60); 6. 7.5 cm tank shell red (7.5-cm KWK L/24); 7. 7.5 cm tank shell 39 (7.5-cm-KWK L/43 & L/48); 8. 7.5 cm tank shell 39/42 (7.5-cm-KWK 42 L/70); 9. 8.8 cm tank shell 39 (8.8-cm-KWK L/56); and 10. 8.8 cm tank shell 39-1 (8.8-cm-KWK L.71).

Between January 1944 and March 1945, 489 of the Panzerkampfwagen VI (Sd/Kfz. 182) "Tiger" II were completed. In a comparable time period, over 1,500 "Stalin" tanks were built for the Red Army. The first new "Tigers" saw action with the 1st Company of Heavy Tank Unit 503 in Normandy. As the picture shows, these tanks used the "Porsche" turret. (BA)

A "Tiger" II with the so-called production turret (Henschel), seen in Budapest with Heavy Tank Unit 503 in October 1944. The 8.8 cm Tank Gun 43 L/71 could pierce 132 mm of armor plate at a range of 2,000 meters. In comparison, the 122 mm D-25 T gun of the Russian IS-2 could penetrate 97 mm at that range. (BA)

The use of captured Russian KW 753(r) tanks originally caused problems because of the limited possibility of recovering these heavy tanks. Some KW tanks were made into recovery tanks.

This captured Russian T-70 was used by Assault Gun Unit 276 in the spring of 1944. Such vehicles often lasted only a short time, often being abandoned for lack of spare parts.

This Driving School Vehicle II was used at the Tank Replacement and Training Unit in Leipzig-Borna in 1944. To give the students a clear impression of the visibility from a tank, a panel was added in front of the driver. In other vehicles the upper body was retained, while others were run on wood-gas.

Driver training on Panzerkampfwagen IV tanks in the National Socialist Driving Corps (NSKK).

In March 1944 the replacement army had 106 driving-school Panzer III and 14 assault guns of the same type. When the front approached, the training units were mobilized and saw action with their vehicles, some of which were rearmed.

Late in the summer of 1944, 13 independent tank brigades were formed as a makeshift solution after the great losses in the previous battles; some of them were equipped with "Panther" tanks. This type of structure did not prove itself; firm leadership and supplying, which were necessary in an armored division, were lacking.

This picture clearly shows the extremely long and slim barrel of the 7.5 cm Tank Gun 42 L/70. The Zimmerit coating on the armor plate, to prevent magnetic charges from adhering, was still customary at that time.

1945:
The End

With the "Armored Division 45" operational structure, efforts were made to correspond more closely to the actual use of armored battle groups. As many armored battle groups as possibly should be at hand on the front to add their strength to it. At the same time, the shortage of tanks should be made up. The monthly production was insufficient, a fact that contradicted the prognoses that were still quite optimistic at the end of 1944.

In the Correct Value Program IV/V of November 7, 1944, the production of 250 Panzerkampfwagen IV was planned for April 1945. This type was to be phased out with 50 tanks in June. In the same month, "Panther" production was to hit a high point of 570 vehicles (April 1945: 460). In actual fact, a total of only 446 Panzerkampfwagen IV were built from January to April 1945. There were 443 "Panthers" built, 20 of them in April.

The orders of the Inspector General of the Armored Troops dated March 24, 1945, were followed by "Armored Division 45" shortly before the war ended. There was no longer a differentiation between armored and armored grenadier divisions. The former 1st Armored Grenadier Battalion with its armored personnel carriers left the armored grenadier regiment and joined the new "mixed" armored regiment. Along with the single remaining tank unit, it belonged, along with the self-propelled gun unit of the artillery regiment and an armored engineer company, to the division's armored battle group. Thus, the number of tanks remaining in a division was reduced to about 80 vehicles. In comparison, in 1939 an armored division had 324 tanks, and in 1943 Guderian still wanted to raise the full strength to 400 tanks. Naturally, with such small armored units, no large-scale offensive tasks could be carried out any more.

The variety of vehicle types in the tank regiments, born of necessity, is shown in this picture, taken near Nagybayom, Hungary, in the first half of March 1945. At left is a Panzerkampfwagen IV, at right an Assault Gun III, and in the background a "Panther" tank.

The restructuring was carried out in only a few armored divisions. One was the "Müncheberg" Armored Division. Following a message of March 5, 1945, it was set up hastily. The comment "KSTN PD 44 als Anhalt" had no value except on paper, since only one mixed armored unit was actually on hand. Originally, this was the "Jummersdorf" tank unit, which became the I./Tank Regiment 29 on March 19, 1945. Its first company had 11 Panzerkampfwagen IV, the second had ten Panzerkampfwagen V "Panthers," and the third had ten Panzerkampfwagen VI "Tigers." (In all, 31 tanks were thus available.) The "Tigers" of the 3rd Company, all Type I, took part in the street fighting in Berlin, near the Brandenburg Gate, at the end of April 1945.

The structure of the "Kurmark" Armored Grenadier Division is interesting. The tank regiment had a First Unit with the staff company and four companies, three of them using the Jagdpanzer 38 "Hetzer." It was made of Panzerjäger Unit 1551. The 2nd Unit of the "Kurmark" Tank Regiment had a staff company and three "Panther" companies.

For the "Clausewitz" Armored Division, according to a structure scheme of April 4, 1945, the "Armored Division" 45 guideline was to be followed. But except for 31 assault guns, no new material could be assigned to this unit. A document states: "The material needed for establishment is to be taken away from the units named in Paragraph I. Further material assignments will not take place." This meant that the vehicles in the tank training unit and the "Grossdeutschland" Tank-destroyer Unit were to be taken over.

In the summer of 1944 the Germans still made the attempt to equalize the quantitative inferiority in the realm of tank armament with qualitatively superior material.

In all, 392 Tank Destroyer V (Sd.Kfz. 173) "Jagdpanther" were produced. They were used primarily in Army tank-destroyer units. This "Jagdpanther" was on duty at the Altdamm bridgehead near Stettin at the end of March 1945. At the beginning of April 1945 there were none of these vehicles left on the eastern front; in the west there were 24.

Structure of the "Münchenburg" Armored Division, March 1945
(simplified portrayal)

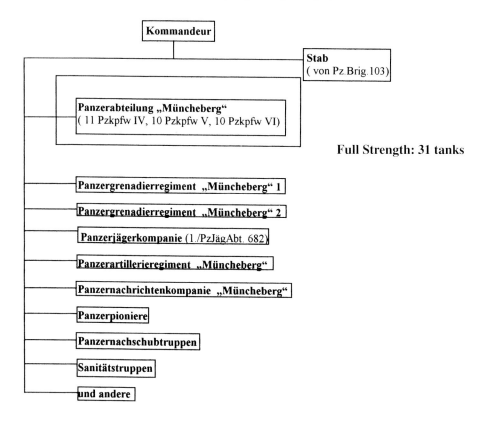

Kommandeur

Stab
(von Pz.Brig.103)

Panzerabteilung „Müncheberg"
(11 Pzkpfw IV, 10 Pzkpfw V, 10 Pzkpfw VI)

Full Strength: 31 tanks

Panzergrenadierregiment „Müncheberg" 1

Panzergrenadierregiment „Müncheberg" 2

Panzerjägerkompanie (1./PzJägAbt. 682)

Panzerartillerieregiment „Müncheberg"

Panzernachrichtenkompanie „Müncheberg"

Panzerpioniere

Panzernachschubtruppen

Sanitätstruppen

und andere

In 1945 all "Tigers" had become rare. In a report on the tank situation early in April 1945, the existence of 65 vehicles was confirmed, with 29 of them on the eastern front. The "Tiger" company of the "Müncheberg" Armored Division, which was established early in March 1945, then had only seven usable tanks of this type. This picture was taken at the Oderbruch in the spring of 1945.

Noticeable in the variety of armored developments on Germany was the great number of tank destroyers. Two of the finest designs were the already mentioned Jagdpanzer 38 "Hetzer" and the much heavier "Jagdpanther" (Sd.Kfz. 173). Important impulses in the further development were gained by the German armored troops from the eastern campaign, in battle against an enemy that grew more and more overwhelming. The influence from the tank battles elsewhere in the war, especially in France, remained meager. The development of anti-aircraft tanks, forced by the Allied air superiority, deserves to be mentioned.

Table: Tank Production, 1944

Soviet Union	USA	Great Britain	Germany
29,000	17,565	2,476	8,466
Total	49,041	8,466	

The most notable further development in the realm of the tank was the "Tiger" II (Sd.Kfz. 182), of which 477 were completed in 1944-45 (18 of them in March 1945). In comparison, the Soviet Union produced 2,250 IS-2 heavy tanks with 122 mm tank guns in 1944 alone.

Still, the new "Tiger" was armed with an 8.8 cm Tank Gun 43 L/71 that was able to pierce 147 mm of armor at 1,500 meters (The IS-2 had 111 mm armor). The bow armor was also noteworthy, reaching a thickness of 150 mm at a 40 degree angle (piercing path about 260 mm), thus offering enough protection. The fighting weight of almost 70 tons naturally inspired discussion within the armored troops about the sense and purpose of such heavy vehicles.

In addition, emphasis was put on the further development of the "Panther," without any definite results emerging by the war's end. Here, too, a tank was envisioned that would weigh around 50 tons. But it was not by chance that lighter fighting vehicles were discussed in the last weeks of the war. The energetic development of midget tanks must be regarded as an extreme case of these discussions. The impossibility of taking the opposite course and building giant tanks had been proved by the building of prototypes like the "Mouse."

As early as the second half of 1943, in view of the growing effectiveness of assualt-gun units while suffering few losses, a basic discussion of the value of assault guns and growing doubt of the effect of tanks took place. Another advantage of assault guns was that they cost less to build. Grenadier units could also cooperate better with assault guns. On the basis of available reports it could be concluded that the tanks of the German armored troops were shooting down fewer and fewer enemy targets. Two examples: In the realm of Army High Command 10 in Italy, 432 tanks had been shot down in May 1944. Of them, 47.5% were scored by tank-destroyer and assault-gun troops and only 13% by German tanks. Of course, it can be said that the Italian area had particular disadvantages for the use of tanks. But very similar results came from a comparable evaluation of the eastern front. The staff officer for tank engagement (Stopa) in Army High Command 9 noted the tanks shot down by various weapons in a report of March 10, 1945: Of 79 tanks shot down in February 1945, only three were scored by tanks, but 31 by tank destroyers and assault guns. Another 30 were destroyed by close-combat means.

The causes of the decreasing effectiveness of the armored troops in battle against enemy tanks can be found in the meager strength of the units, the shortage of experienced tank crews, and surely also in the tactics of the tank units. The fact that the tank soldiers' morale was no longer very high is shown by orders from the spring of 1945. In cases of going about tasks frivolously, the tank crews were subject to the sternest investiga-

The armored divisions participating in the fighting in the west at the end of 1944, including the 116th, the tank training division, and various SS armored divisions, had been newly reorganized. In terms of weapons, the "Panther" was superior to almost all British and American tanks. (BA)

A Panzerkampfwagen V (Sd.Kfz. 171) "Panther", Type G, drives through a passage in a tank barrier of the West Wall in the winter of 1944-45. (BA)

tion and punishment. Still, in all, the actual action strengths remained meager. Information is provided by a message from February 14, 1945, concerning the Führer Grenadier Division:

Type	Full	Ist	Ready for action
Assault Gun III (Sd.Kfz. 142/1)	45	33	24
Assault Gun IV (Sd.Kfz. 163)	?	12	7
Panzerkampfwagen IV (Sd.Kfz, 161/2)	28	13	1
Panzerkampfwagen V "Panther" (Sd.Kfz. 171)	36	14	3
Anti-aircraft Tanks	8	4	1
Total	117	66	36

Barely one third of the full strength was ready for action, and again the high percentage of assault guns is of interest. In the last days of the war, the combat strength often decreased considerably for lack of supplies.

Often in pocket battles, as in the Halbe area at the end of April 1945, it was small armored battle groups, often only individual vehicles, around which the still viable units gathered to carry on the battle.

In city street fighting, tanks, as already noted, could utilize their mobility and firepower only in a very limited manner. But these were precisely the characteristics that made it difficult to form focal points of defense, what with the widespread lack of heavy weapons. Thus, tanks were put to use there. This was true of the final combat in Berlin and other cities. Naturally, this form of combat had nothing to do with the basic functions of the armored troops practiced so successfully at the beginning of the war.

The Wehrmacht's tank situation at the beginning of April 1945 shows 1,655 tanks (not counting 230 captured tanks) and 2,084 assault guns (not counting 218 captured ones). This ratio reflects in material terms the significant change in the German conception of the use of armored troops during the course of the war. In providing weapons, the defensive components clearly dominated, for tank technology and the basis for their use had changed.

In the first war years, the German armored troops were the decisive means of achieving weighty, ground-gaining attack operations. At the end of the war, their roles changed in many ways; they were supposed to help hold off the Allied forces. Thus, they had no hope of success.

This Panzerkampfwagen V (Sd.Kfz. 171) "Panther," Type G, belonged to Tank Brigade 107 and was hit by a PIAT shell near Overloon on October 13, 1944. The tank can be seen today in the Overloon Museum Park of the Netherlands War and Resistance Museum.

In the 9th Army's breakthrough battles in the Halbe area at the end of April 1945, the few Panzerkampfwagen VI (Sd.Kfz. 182) "Tiger" II tanks of the Heavy SS Tank Unit 502 played a decisive role in several fights. Early in April the unit still had 32 tanks (of which 27 were ready for action).

By rebuilding 56 B-IV heavy charge carriers, Tank Unit (FKL) 303 created a small tank destroyer armed with six 8.8 cm Rocket Antitank Gun 54 weapons united in a block of six barrels. The vehicles were used by Tank Destroyer Unit 1 in Berlin at the end of April 1945.

The Allies' absolute air superiority on the western front made maneuvering with even small armored battle groups difficult. Fighter-bombers attacked these Type G "Panthers" and put them out of action. (BA)

This Panzerkampfwagen IV (7.5 cm L/48)(Sd.Kfz. 161/2), Type J, of the 20th Armored Division was left near Altenberg in the Erzgebirge in May 1945. Note the white-bordered triangle on the turret's front skirt armor.

This "Panther" was lost in Posen in February 1945 after taking several hard hits.

Six of the 15 cm Heavy Infantry Gun 33/1 on Self-propelled Mount 38(t)(Sd.Kfz. 139/1) saw service in each infantry gun company of the armored grenadier regiments, where their main task was supporting the armored grenadiers who operated along with the tanks. There were 282 of them made. This vehicle was abandoned in Müglitz in the Erzgebirge at the war's end and photographed there in 1946.

For the beginning of April 1945, the Inspector General of the Armored Troops listed a total of 538 Panzerkampfwagen V (Sd.Kfz. 171) "Panther" tanks. Only 299, a little more than half of them, were ready for action. The high losses could no longer be made up because of decreasing production.

Among the tanks that were assembled in Glashütte by the Red Army for transport was this Panzerkampfwagen V (Sd.Kfz. 171) "Panther," Type D., production of which had begun in September 1943. It was seldom that a tank remained in service for so long. Obviously it belonged to a training unit.

Foreign workers on the way to their homelands pose on this Panzerkampfwagen I of the 10th SS Armored Division "Frundsberg." This tank was abandoned in Glashütte on May 8, 1945. The vehicle shows typical signs of the final version, including the road wheels without rubber tires and the simplified attachment of the track aprons.

The end of the once-proud armored troops on a scrap heap at the Lauenstein railroad station in the Erzgebirge in the summer of 1945. The exhaust mufflers, typical of the Type J Panzerkampfwagen IV, can be seen clearly.

The 7.5 cm Tank Gun 42 L/70 ranked in its time among the most powerful tank weapons. The gun, firing a Tank Shell 39/42 striking at a 60-degree angle at 2,000 meters, had a penetrating power of 88 mm of armor plate (Vo 925 m/sec). With the Tank Shell 40/42 under the same conditions, 106 mm could be pierced. The barrel of the 7.5 cm Tank Gun 42 L/70 was 5,250 mm long (= L/70); the whole gun weighed some 1,500 kilograms. The 7.5 cm Tank Shell 39/42 reached a velocity of 935 m/sec, the 7.5 cm Tank Shell 40/42 attained 1,120 m/sec.

This Type G "Panther" was left in Lauenstein in the Erzgebirge in May 1945. The road wheels were taken off by the people and used as wheels for trailers, and some of the rubber tires were used to make shoe soles.

Calling the "Mouse" tank the high point of German tank construction in World War II would be based only on its weight of almost 200 tons. The two prototypes were at the Army Test Center for Tanks and Motorization in Kummersdorf. From there, drivers from this base were supposed to drive the tanks to Zossen to protect the Army High Command there. Both vehicles were abandoned, one being blown up, and...

the other falling into the Red Army's hands with minor mechanical damage. This picture shows the Russians "capturing" the Mouse.

The "Mouse" tank can be seen today in Kubinka, near Moscow.

A few Panzerkampfwagen IV with short-barreled guns had survived as driving-school tanks and saw action in Germany at the end of World War II. This turret, found near Gotha in Thuringia, came from such a vehicle.

Some 160 to 170 tanks and assault guns were still with the 4th Tank Army that operated in Saxony at the beginning of May 1945. As the first Ukrainian front began its offensive in the direction of Prague on May 6, it could make use of 1,600 in all. After the war, Russian occupation troops gathered the abandoned German tanks as spoils of war. This picture was taken in Glashütte, Saxony, in the summer of 1945.

Now and then, remains of German tanks destroyed in the Erzgebirge in 1945 are still found. The reconstructed turret of a Type G "Panther," with a reinforced roller mantlet and deflector panel on the turret roof, was one of them; it can be seen in the Museum of Military History in Dresden.

Bibliography

MZA Potsdam: SF 01/3959, SF 03/16727, WF 03/3428, SF 03/35224, WF 10/2432, WF 10/2433, SF 10/2436, WF 10/2499, WF 10/12749, WF 10/13433, WF 10/13676, etc.

H.Dv.470/5: Training Directions for Armored Troops. Training for the Tiger Tank, 8/15/43, Berlin 1943.

H.Dv.470/10: Temporary Guidelines for Command and Combat of the Tank Regiment and Tank Unit, Berlin 1941, etc.

D 651/12: Panzerkampfwagen II (2 cm)(Sd.Kfz. 121), Types D and E, of 4/15/39, Berlin 1942.

D 652/52: Panzerkampfwagen III, Types L, M, N, of 5/15/43, Berlin 1943.

D 653/7: Panzerkampfwagen IV, Types F 1 and F 2, of 4/1/42, Berlin 1942.

D 655/1b: Panzerkampfwagen Panther, Types A, D and variants, of 7/21/44, Berlin 1944.

D 655/27: Panther Manual of 7/1/44, Berlin 1944.

D 656/27: Tiger Manual of 8/1/43, Berlin 1943.

D 736/1: Equipment Description and Service Manual for the upper body of Panzerkampfwagen I (MG)(Sd.kfz. 101) of 4/1/36, Berlin 1939.

Ebeling, K.: Written report on the 9th Army's escape from the Halbe pocket, Gelsenkirchen-Buer 1991.

Lehmann, A: Oral report on Tank Regiment 33, Egsdorf 1984.

Borchert, H. W.: Panzerkampf im Westen, Berlin 1940.

Carisius, O.: Tiger im Schlamm, Neckargemünd 1960.

Fechner, F.: Pamzerkampf im Westen, Berlin 1940.

Kaufmann, K.: Panzerkampfwagenbuch, Berlin 1939 and 1940.

Kleine & Kuhn: Tiger—die Geschichte einer legendären Waffe, Stuttgart 1984.

Klose: Kriegstagebuch der 4. Kompanie Panzerregiment 35, no place or year.

Lehmann, R.: Die Leibstandarde, Vol. I-V, Coburg 1993ff.

Anonymous: Orel, Die julischlacht 1943, Moscow 1943.

Oswald, W.: Kraftfahrzeuge und Panzer der Reichswehr, Wehrmacht und Bundeswehr, Stuttgart, various editions.

Popjel, N. K.: In schwerer Zeit, East Berlin 1962 and 1981.

Scheibert, H.: Kampf und Untergang der deutschen Panzertruppe 1939-1945, Friedberg 1992.

Strauss, D.F.J.: Friedens- und Kriegserlebnisse einer Generation, Neckargemünd 1977.

Wittek, E.: Die soldatische Tat, Berlin 1941 (selected)

Periodicals:
Deutsche Wehr
Die Panzertruppe (ex-Kraftfahrkampftruppe)
Die Wehrmacht
Militärgeschichte
Militär-Wochenblatt
Wehrtechnische Hefte
Zeitschrift für Heereskunde